TACTICAL RUGER 10/22

To Peder, thank you for this opportunity and the kindness you have shown
to my family and me over the years. We will forever be grateful.

TACTICAL RUGER 10/22

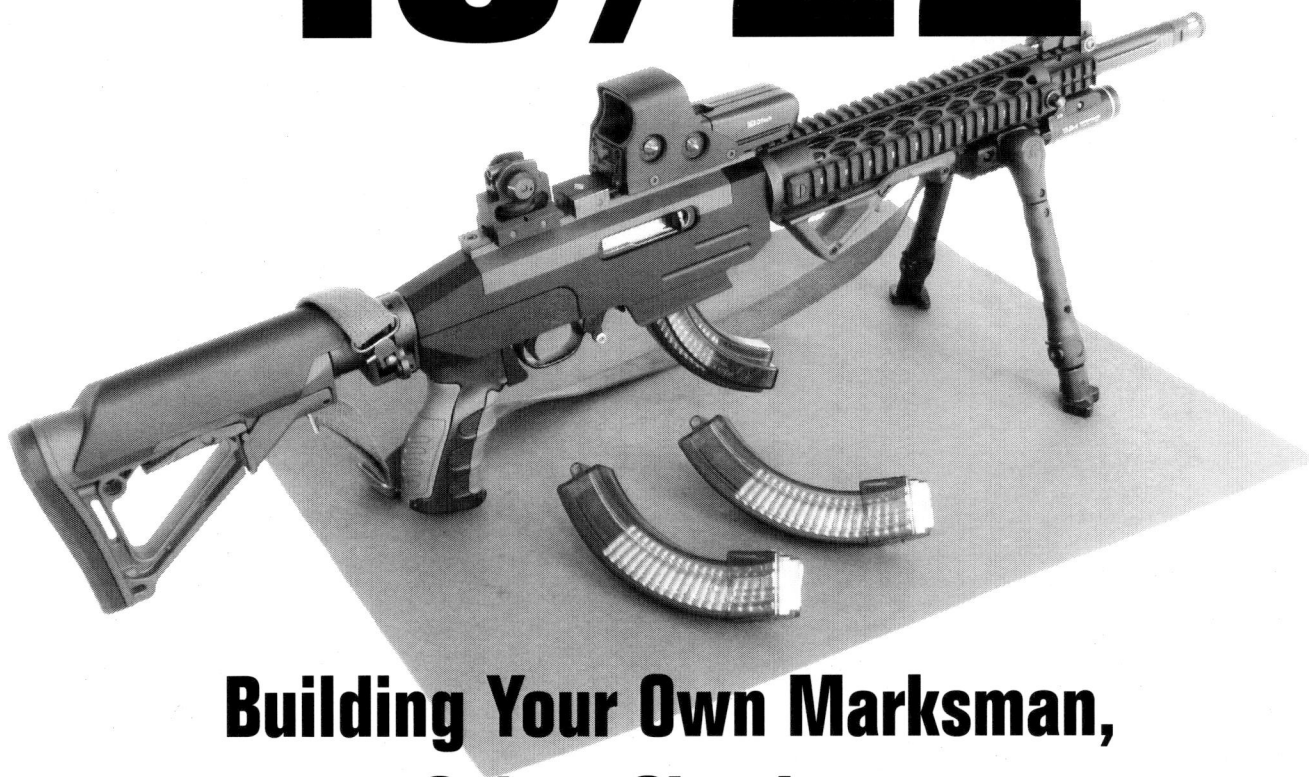

Building Your Own Marksman, Sniper Simulator, and Competition Models

J.M. Ramos

PALADIN PRESS • BOULDER, COLORADO

Tactical Ruger 10/22:
Building Your Own Marksman, Sniper Simulator, and Competition Models
by J.M. Ramos

Copyright © 2012 by J.M. Ramos

ISBN 13: 978-1-61004-573-5
Printed in the United States of America

Published by Paladin Press, a division of
Paladin Enterprises, Inc.
Gunbarrel Tech Center
7077 Winchester Circle
Boulder, Colorado 80301 USA, +1.303.443.7250

Direct inquiries and/or orders to the above address.

PALADIN, PALADIN PRESS, and the "horse head" design
are trademarks belonging to Paladin Enterprises and
registered in United States Patent and Trademark Office.

Visit our website at www.paladin-press.com

Table of Contents

Introduction

The second decade of the 21st century is shaping up to be a very exciting era for a new breed of firearms called "tactical .22s." This new generation of rimfire clones can be best described as a modern resurrection of the once very popular rimfire look-alikes of the 1980s. The first-generation .22 clones, primarily those of European origin, were of high quality and remain authentic looking even by today's standards. On the other hand, the latest versions of these militarized rimfire self-loaders are far more elaborate and sophisticated than their vintage predecessors. What truly makes these new radical .22s interesting and extraordinary is the fact that they are now made by the big names in the military arms industry and not just sporting arms manufacturers as in the past. Many of the factory-produced tactical .22s share furniture and accessories available for their full-bore military weapon counterparts, making them virtually indistinguishable from one another, except for the magazine's external profile.

Perhaps the most interesting development between the 1980s exotic .22s and today's tactical-inspired superclones is the way they came into being. Back in the 1980s, the Cold War between the world's two superpowers was at its height. It was the era of the politically incorrect "assault rifle." The infamous Soviet AK-47 and U.S. M16 rifles seen in combat graced the front pages of daily newspapers, as well as appearing in television footage and such movie blockbusters as *Rambo, Red Dawn,* and *Commando.* These high-octane war movies showcased the latest military hardware of the time.

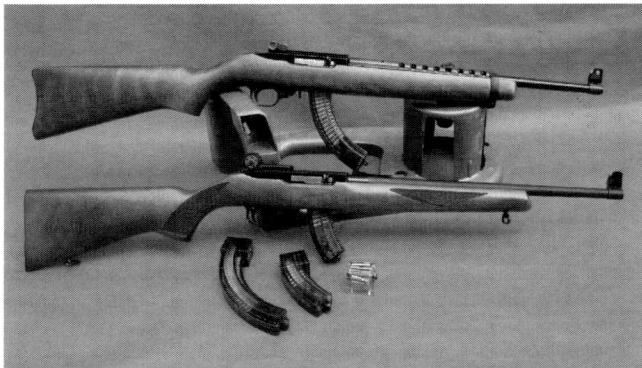

Prime examples of vintage tactical 10/22s upgraded by Tech-Sight's TSR100 GI-style military sight sets and Ram-Line's high-capacity banana-type magazines.

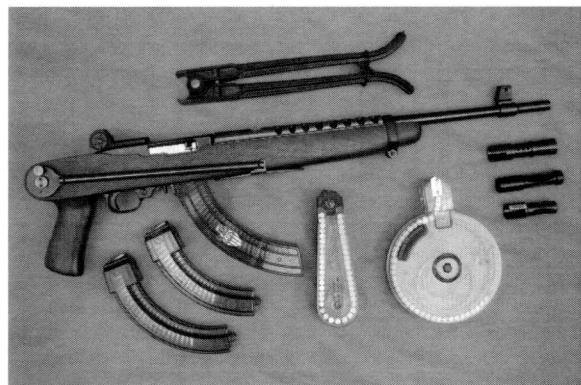

Another classic tactical 10/22 of the 1980s is this carbine weapon system featuring a Federal Ordnance German MP-40 style under-folder stock and various high-capacity magazines made by Eagle International, Ram-Line, Sanford, and Mitchell Arms. Note the clip-on plastic bipod and early slip-on flash hiders.

The real stars during the paramilitary heyday of the Reagan years were the Jager .22 rimfire clones made in Italy. The FAMAS .22 (top) and Galil .22 (bottom) are prime examples of these early-generation tactical .22s equipped with high-capacity magazines, scope mounts, and folding steel bipods.

The sky-high cost of centerfire ammunition has put a damper on training for the busy marksman. The only viable alternative is the .22 LR cartridge. Today's custom tactical chassis for the 10/22 allows replicating the handling characteristics of the military M4/M16 rifles, making it the ideal training apparatus for military weapons enthusiasts. A tactical 10/22 is also a perfect training companion for a 1911 pistol with .22 conversion kit in IPSC rig.

These classic military battle guns, especially the hard-to-get Soviet-made AKs, remained in high demand among diehard military arms collectors and shooters from the 1960s up through the 1990s. This surge of interest in military weapons among civilian enthusiasts eventually led to the creation of the first generation of rimfire clones, including the Italian-made Jager M1622, the AK47 .22, Galil .22, FAMAS .22, Bingham PPS-50, U.S.-made Thompson .22, Erma M1 .22, and others. Today, these vintage rimfire clones command top dollar by collectors when found in mint condition.

On the other hand, the initial introduction of today's tactical .22s can be attributed to three basic factors: (1) the worldwide appeal of military/police tactical-type weapons in the civilian market; (2) the election of Barack Obama in 2008 that triggered panic buying among military weapon enthusiasts, who feared that passage of another gun-control bill would once again prohibit this class of firearms; and (3) the unexpected shortage of centerfire ammunition for full-bore tactical, police, and military firearms.

The worldwide economic decline also contributed to the rapid introduction of tactical 22s. As the economic chaos that started at the end of 2008 spread around the globe, arms manufacturers soon realized that the only way to survive in the business was to come up with an appropriate product line to

deal with this unusual problem of high ammunition costs and shortages, something the gun industry had not experienced before. This led to the creation of the many tactical .22s we see on the market today, produced by such firearm luminaries as Smith & Wesson, FN, SIG, Heckler & Koch, and Ruger, among others. Although Colt Firearms has the most models in this weapon category, its versions are not made by the company itself, but rather by the famous German arms maker Walther GmbH.

While the lineup of elite military and police arms manufacturers that jumped on the tactical .22 bandwagon is unprecedented, German Sports Gun (GSG) is generally credited with being the first to initiate this new breed of .22 superclones with the introduction of its superb GSG-5, a faithful rimfire replica of the famous H&K MP5 submachine gun (available in carbine and pistol versions). This model was soon followed by another authentic-looking clone, the GSG AK-47 .22, a rimfire copy of the classic Kalashnikov battle rifle, available in fixed- and folding-stock versions.

Another class of tactical .22s has been causing a sensation among tactical firearms aficionados and hobbyist gunsmiths alike for quite some time. Popularly known as "rail guns," this custom tactical

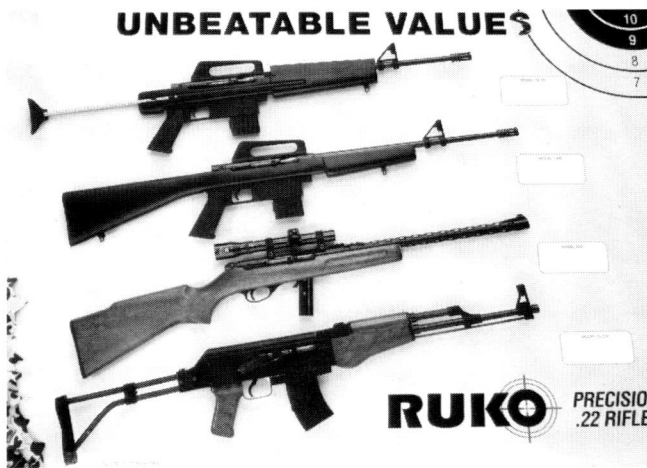

A quartet of economically priced Philippine-made .22-caliber rimfire clones of the M16, AK-47, and PPS-50 SMG as imported in North America in the 1980s. The AK-22 and M16 .22 are still produced today but have seen limited importation.

.22 utilizes a special platform or chassis system to accommodate the action of the gun. It is then completed to the desired configuration of the builder using a host of aftermarket accessories and furniture.

The basic idea of accessorizing certain models of sporting-type firearms has been around for more than three decades. The most popular among the best-selling models is no doubt the Ruger 10/22, which just celebrated its 5-million production mark not too long ago. Since the 1980s, the 10/22 has seen amazing custom upgrades, more than all its competitors combined. From folding stocks to high-capacity magazines and everything in between, America's independent accessory producers have not yet run out of money bettering Bill Ruger's self-loading rimfire carbine.

The 1980s exotic gun market was dominated by European look-alikes. Among the many popular U.S.-made rimfire self-loading rifles available at the time, only the 10/22 really gave these sophisticated imports a run for their money. The 10/22's reign as the king of U.S. rimfire self-loaders, as well as its ability to compete with the European models, was made possible by the introduction of independently produced aftermarket accessories, such as folding stocks, high-capacity magazines, hand guards, slip-on muzzle devices, clip-on bipods, and quick-detach (QD) scope mounts.

These vintage bolt-on gadgets undoubtedly inspired the latest in state-of-the-art accessories, designed to meet the demand for a tactically designed 10/22 among the new generation of shooters and military gun buffs. Typical dress-up kits of the past consisted of a pistol grip stock with fixed or folding butt complemented by a metal or plastic ventilated hand guard. The setup was then equipped with a slip-on flash hider or muzzle brake along with a clip-on plastic bipod. Other innovative additions included a QD-type scope mount, normally affixed on top of the receiver, and rudimentary military-style replacement iron sights, such as the Zypher sighting system made by Eagle International (available for the 10/22 and AR-7). The Zypher can be considered an early version of a backup iron sight, which is so popular on today's tactical rifles and shotguns.

Among the many pioneers of aftermarket dress-up kits during the Reagan years, Ram-Line truly revolutionized early weapons accessorizing with its many innovations, such as the all-polymer AR-15 kit, which consisted of an AR-style carry handle mounted to the gun's factory scope mount and a matching AR-style front sight tower that rode over the factory front sight using a plastic fastener. Complementing this kit was a straight-stock plastic folder, matching ventilated hand guard, and birdcage-style flash hider made of aluminum. Perhaps the most remarkable invention by Ram-Line was the double-stack, high-capacity magazine in banana-type configuration, available in compact 50-, 40-, 30-, 20-, and 15-shot capacity.

Other high-capacity magazines developed and marketed for the 10/22 during that time that have been resurrected today include the Sanford 50-shot drum magazine (manufactured in slightly different format by Pro-Mag and Black Dog). MGW now produces the Mitchell 50-shot teardrop-style magazine. Butler Creek and a host of other companies now produce improved versions of the original 25-shot Eaton magazine with plastic and aluminum body construction. The original 30-shot Eagle magazine is produced today by a new Eagle Company and is available in a variety of colors. Choate Machine and Tool still manufactures its classic fixed and folding stocks for the 10/22 but with a shorter buttstock.

The latest tactical weapon accessories are a far

cry from their 1980s counterparts. Although the basic principle remains the same, the manufacturing methods and the types of components used today are totally different. Today's high-tech enhancements provide the builder with unlimited options for weapon configurations, limited only by his expertise and imagination. This decade is truly a paradise for the tactical gun builder.

The Nordic custom AR chassis system for the 10/22, introduced just a few years ago, caught an eye at Ruger's research and development department, and the company eventually selected it for its new SR-22 tactical carbine. The sensational SR-22 is Ruger's answer to the ever-growing tactical .22 lineup offered by Colt, Smith & Wesson, FN-USA, SIG SAUER, and Heckler & Koch, as well as other AR rifle–producing companies that opted to utilize regular .22 conversion kits (such as the Atchisson or Foote units) instead of coming up with a new design to compete with this new breed of firearm.

What makes the new Ruger SR-22 extra-special is its Nordic chassis system. This custom-made component is designed to take any AR stocks (fixed, M4 style, or folding), pistol grip, forearms (one-piece tubular or the two-piece type using the usual AR-15 Delta ring assembly and hand guard cap), backup iron sights, and rails. Other competing models have a cast-in buttstock extension and, therefore, will not accept folding-stock variants, custom barrels, or fixed or noncompatible forearms with mil-spec parts and accessories. This limits their capacity to be transformed into other weapon profiles using the same action. You would have to purchase virtually all the models a company makes in order to have all the desired configurations, something the SR-22 platform can offer with the same basic action by simply substituting AR parts and accessories, including varieties of folding stocks for those who want a more compact configuration for their tactical 10/22s.

Another advantage offered by the SR-22 or other 10/22s that utilize custom tactical platforms is the abundance of match trigger mechanism modules, high-capacity magazines, custom barrel selections, and everything in between. Competing models are normally offered with 25- or 30-shot magazines, whereas the 10/22 can be had with up to 50-shot

Two custom-upgraded tactical 10/22s featuring an early Ram-Line wood/metal side folder and Choate Zytel folding stock. The top gun is a transitional hybrid tactical 10/22 utilizing the Krinker-Plinker rear and front sight, complemented by a Choate Mini-14 flash hider. The bottom gun has an M1 .30 carbine rear sight and Tech-Sight front sight. The flash hider is an original M16 XM177 Commando.

A beautiful fixed-stock "compact model" 10/22 in tactical format. Accessories include Krinker-Plinker iron sights, CAA VG1 forward grip, modified Mini-14 flash hider, and Harris bipod mated to a CAA adapter. The optic is a Bushnell Trophy red dot. The "quad" 50-shot Ram-Line magazine provides awesome firepower.

capacity in banana, drum, or teardrop configuration. For decades this gun has proven itself to be a winner no matter what the challenge. And today's tactical .22 competition seems no exception.

Building your own tactical 10/22 has never been

INTRODUCTION

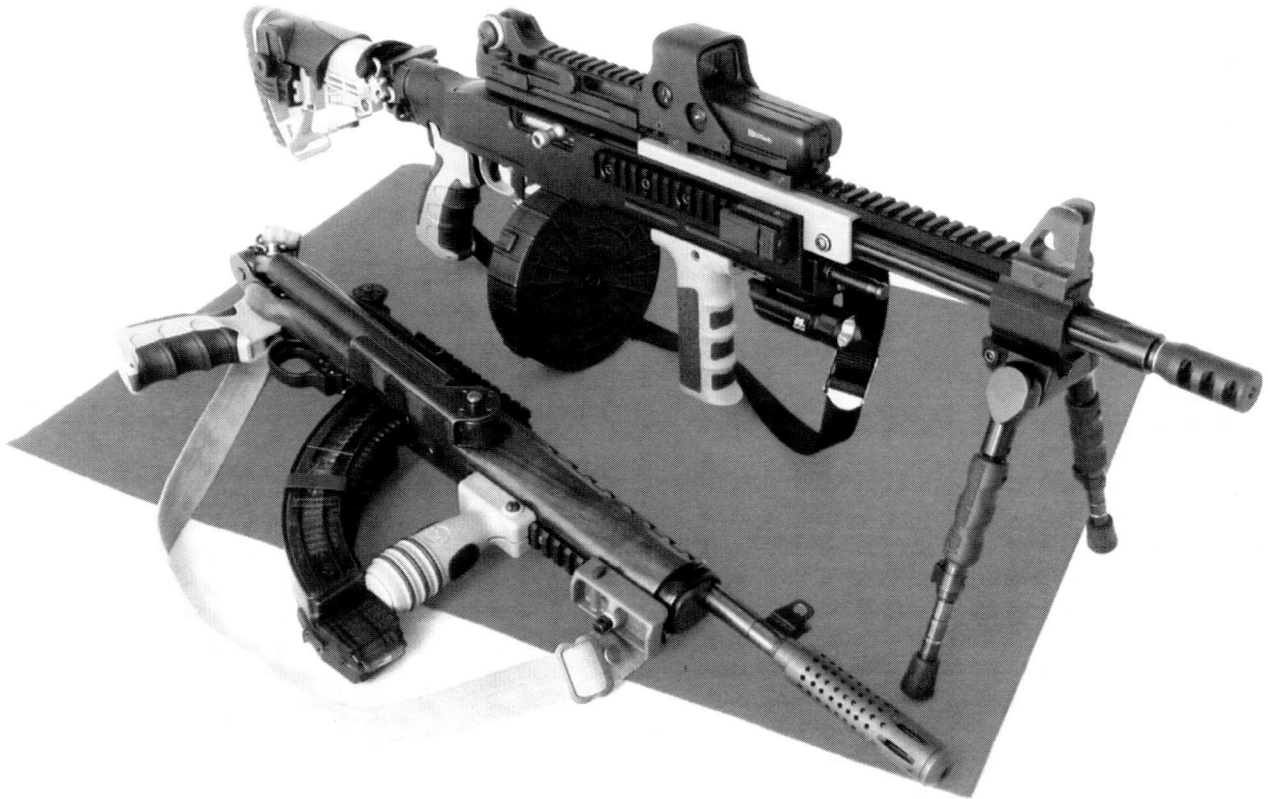

Comparison between the 21st-century tactical 10/22 (top) and its vintage counterpart slightly upgraded with CAA pistol grips and bipod adapter (bottom).

easier. The art of tactical gunsmithing and customizing is sweeping the gun world by storm, be it long or short guns. As the quest for the ultimate tactical firearm rages on among the elite producers of military hardware, so too does the competition among independent accessory innovators, which allows the creation of even more astonishing customized tactical .22s.

A couple of decades ago, it was thought that the 21st century would be the era of star wars and ray guns. The truth is, the foot soldier of today is still fighting with Grandpa's vintage burp guns. The only difference? We have simply made them look and perform better.

Note: Some of these rifle configurations may require special permits to build in certain jurisdictions. Check all applicable state and municipal regulations before construction.

CHAPTER 1

From Fantasy to Reality

Every gun lover has at one time or another fanta-sized about owning the gun of his dreams, with all the bells and whistles specially designed to meet his particular preferences and requirements. Wishing is a good thing, but making it happen is even better. Now you don't have to just dream of owning the perfect gun. Tactical gun aficionados can make their own dream guns, just the way they want them.

The development and introduction of the innovative tactical weapon rail system, which has now been uni-versally adapted to almost every type of combat rifle, is revolutionizing the weapons market. New-generation military arms are being produced with built-in rails to accept a multitude of combat accessories. Now, tactical guns are being fashioned after these modern high-tech war machines. The designs of the new factory-made tactical .22s are derived from these marvelous fighting arms as well as their custom-built versions that utilize special platforms.

This book is intended as a hands-on guide for the gunsmith and firearm hobbyist alike. Building your tactical dream gun is easier than ever before, with the aid of these modern bolt-on accessories to dress up your favorite handgun, rifle, or shotgun. Whether it's a rim-

You can tailor your tactical dream gun in any size or for-mat you wish by using the appropriate accessories and chassis. The superb carbine (top) and PDW (bottom) were created using RB Evo and FS556 chassis (respectively) and complemented by a host of genuine AR-15 furniture and accessories.

This marvelous all-purpose tactical rifle was created using the Nordic AR-22 chassis complemented by hosts of AR-15 furniture and accessories. Nordic has so far created the most versatile 10/22 tactical platform that can transform the 10/22 to any configuration the builder may desire.

The PSG22 rifle is the epitome of a .22 sniper simulator. This magnificent setup comes complete with battle-proven EOTech red-dot scope, laser, light, bipod, backup iron sights, 50-shot drum magazine, and two-tone Magpul furniture. It doesn't get any more authentic than this!

The Canadian-made FS556 is a force to be reckoned with in the 10/22 tactical chassis competition. This superb short-barreled specimen says it all. The top gun features an ACE side-folding stock while the bottom gun sports an M4-style collapsible stock in an eye-catching two-tone design.

fire or centerfire tactical gun that you are planning to customize, the basic idea is the same.

Tactical customizing is unlike custom pistol-smithing, where you have to build the gun from the ground up. Pistolsmithing may involve extensive machine work, from checkering to proper mating of the slide and frame to fitting of critical trigger mechanisms to ensure safe and reliable operation. Tactical gun-smithing is easier to accomplish. It requires a minimum of tools (except for barrel or receiver modifications). Gun dressing is also safer to accomplish since no mechanical alteration to the firearm is required. The whole process is also very exciting simply because it allows the builder to easily change the configuration of the weapon to virtually any format desired by simply changing the furniture, barrel, and other accessories. It's like having a new gun all the time.

The availability of the new-generation chassis systems combined with genuine mil-spec furniture finally made it easy for anyone to create virtually any AR-type weapon out of the 10/22 action. The possibilities range from sniper/competition models to M4-type carbines to compact Class 3 short-barreled rifles (SBR) or pocket-sized PDW/micro variations. Once you have learned how to properly select and coordinate these high-tech accessories, you will be amazed how easy it is to create a personalized multi gun weapon system by simply switching exterior

The ultimate 10/22 combination can be built using the Nordic AR-22 chassis in this very colorful two-tone accent in carbine (top) and PDW format (bottom).

components and related accessories. Building a high-tech rail gun is easy and fun. The benefit to you may extend beyond satisfaction with your custom 10/22 rifle. It could also be expanded to a financially rewarding career for those who decide to offer their newly acquired know-how to collectors and shooters who prefer not to assemble their own tactical weapons.

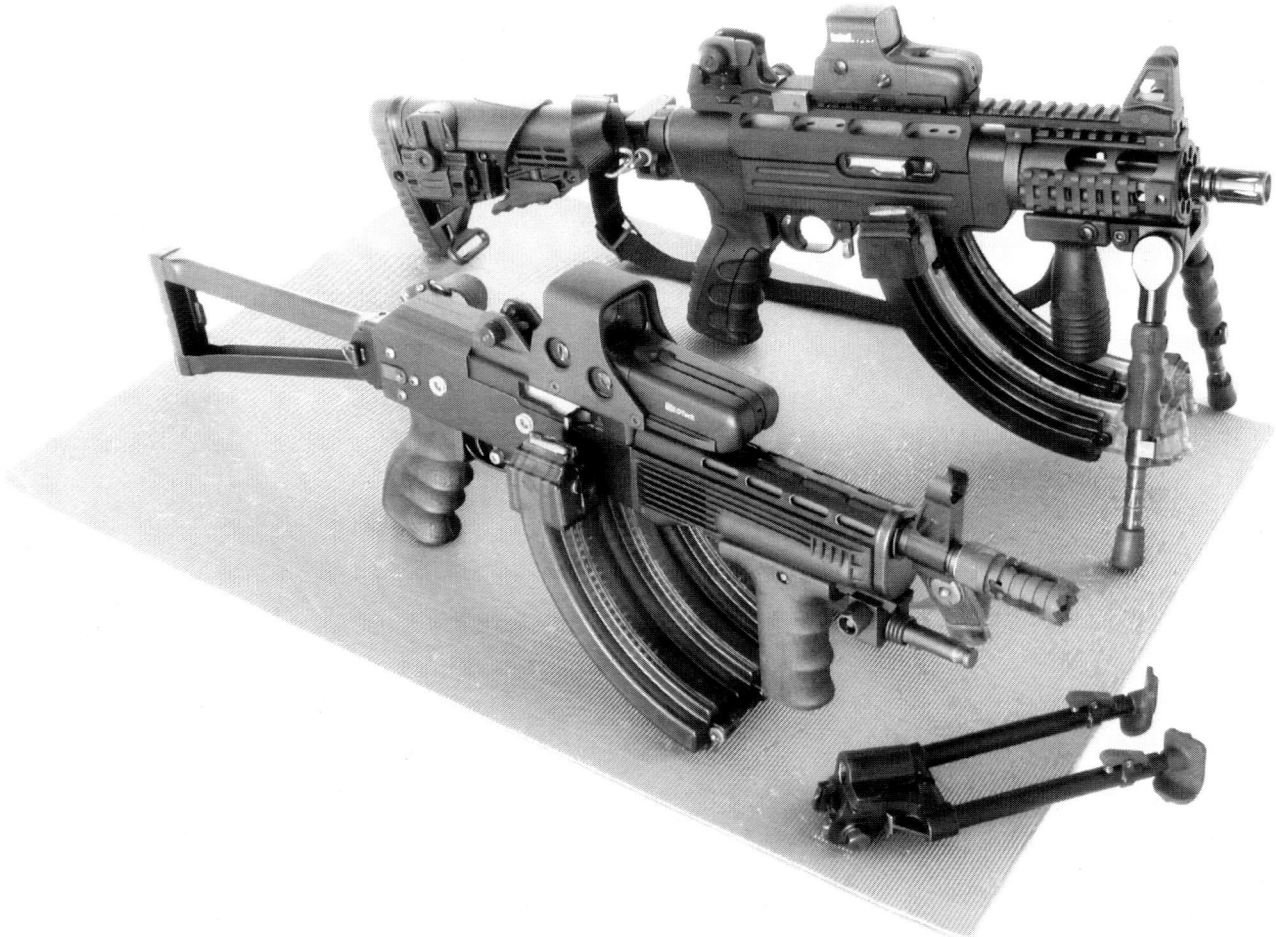

The ultracompact AR-22 PDW (top) compared to Krinker-Plinker short-barreled rifle (SBR). If your preference is a 10/22 in PDW format, nothing smaller than these tantalizing specimens is available. *Caution:* In the United States, this weapon class requires a special license and a $200 transfer tax fee. Also check state and local laws before attempting to build any short-barreled rifle.

This sophisticated 10/22 PDW-K format fits inside a briefcase with the stock in the folded position. It is a perfect rimfire training companion for the H&K MP5K SMG for licensed government agencies and VIP-protection specialists.

CHAPTER 2

Choosing the Right Platform

Many interesting tactical dress-up kits are available for the 10/22. These accessories vary in style, quality of workmanship, and price. Smaller independent companies, such as Rhineland Arms and Dixie Consolidated, were some of the first entrants in the 10/22 tactical chassis competition, along with RB Precision (Evolution stock) and CSM Werkes (Krinker-Plinker).

Rhineland came up with several unique and interesting setups for the 10/22, including the MP5SD-style kit with an H&K-style, side-mounted, collapsible metal stock complete with a suppressor. An ultra-compact version of this stock kit—a mere 16 inches overall, measured from the rear of the stock to the front of the hand guard—was made in prototype form. This short version was slated for registered Class 3 10/22s with short barrel and suppressor. However, this ultracompact model did not reach the production stage. Rhineland managed to produce a limited run of its 10/22R series of stocks featuring the company's trademark stock kit with side-mounted collapsible stock and H&K-style ventilated forearm. The first-production CNC-machined stock from Rhineland is very similar in concept to the popular RB Precision Evolution stock without the matching hand guard.

Dixie Consolidated focused its design on converting the 10/22 into a bullpup configuration, utilizing a single tubular chassis system to enclose the 10/22 action. The arrangement is quite sound and simple, but it lacks the aesthetic appeal of a modern tactical arm so revered by today's military gun buffs. Regrettably, Dixie Consolidated has closed its doors, possibly due to the recent economic downturn or simply because it could not compete with better designs offered by more stable companies.

A foremost California-based military arms stock maker, Ironwood Design, also came up with several interesting dress-up kits for the 10/22 made of exotic hardwoods. These include a very sophisticated Walther 2000 sniper-type kit and Steyr AUG bullpup-style rifle. Fortunately, this company continues to cater to more traditional audiences who prefer wooden furniture to plastic and metal combinations.

Also on the market for quite some time now is the popular Krinker-Plinker kit. This AKS-74U inspired dress-up kit is made by CSM Werkes and is available in two versions: a short-barreled rifle (SBR), or mini-"Krinkov," and a full-size model with a 16-inch barrel and a fake suppressor that encloses the front end of the barrel to mimic a short-barreled gun with a suppressor. The Krinker-Plinker kit is a great design, well made and quite authentic looking. For 10/22 owners who like the AK look, this kit is unbeatable.

The most recent offering for a 10/22 dress-up kit comes from the E. Arthur Brown Company, the M1 Carbine 10/22 Tribute rifle. The kit consists primarily of a two-piece M1 Carbine-style wooden stock complete with authentic GI oilier and sling. The military-style sights are designed and produced by Tech-Sights, a foremost manufacturer of high-quality, GI-style replacement sights for the 10/22, SKS, AK-47, and Marlin .22 rifles.

For AR rifle fans, there are three superbly crafted chassis systems available today. These include the RB Precision Evolution stock, the Nordic AR-22, and the Canadian-made Fabsports FS556. Each of these kits has its own particular styling and characteristics. Each costs about the same, and all are CNC-machined (computer numerical control, which entails the use of milling heads to cut away unwanted material) from aluminum billets, which are then hard coat–anodized to military specifications, making them first rate in quality and ideal as a starting point for those wanting to build their ultimate 10/22 tactical rifle.

This book presents different weapon profiles utilizing all three superlative platforms in ultramodern tactical formats. I have spent many years creating various tactically inspired 10/22s, using various kits and accessories dating from the early part of this decade up to the very latest in tactical gadgets available today. Models featured in the book are the result of my experimentation and testing. During my research and development work, I discovered several interesting facts. I found out that with the current plethora of innovative products geared toward tactical firearms, the most expensive brands are not necessarily the best ones to use, especially in a .22-caliber gun. There are many affordable accessories that are almost (if not just) as well made as the most expensive brands out there.

The real secret in building a successful tactical 10/22 is choosing the ideal platform and the right accessory combination to go with it, depending on what type of configuration you want. You must be careful to not overdress the gun, as it will become too heavy, bulky, and cumbersome. Keep in mind that if you are using a match-grade bull barrel, you are already starting with a heavy action. Therefore, as much as possible, select accessories that will not add unnecessary bulk or weight to the gun unless it is your desire to build it that way. Full-bore military/police sniper and marksman/competition models are built to virtually the same specifications and standards. Both are typically equipped with a heavy match-grade barrel for superior accuracy, tai-

lored for long-range scope mounting, and sport a multifunction adjustable buttstock and bipod. Some sniper rifles are equipped with a special muzzle brake or suppressor, whereas a competition model may or may not have a muzzle device.

Whatever style of 10/22 tactical gun you have in mind, the models presented herein are well thought out, ergonomically perfect, and accuracy-tested for the purpose for which they have been built. They are presented here to serve as a basis for do-it-yourselves to create their own customized examples of world-class tactical .22s that they own and can use for many years to come.

The following is a brief description of the three basic chassis systems used in the project guns described in the book: RB Precision Evolution, Nordic AR-22, and Fabsports FS556. It is up to you to choose which of these modular components best suits your personal taste and requirements.

EVOLUTION STOCK SYSTEM

The RB Evo stock Gen-1 is a two-piece modular-type platform. The lower half of the stock (main body) is CNC-machined from 6061 T6 aluminum billet. It is hard-coated to a mil-spec black-oxide finish

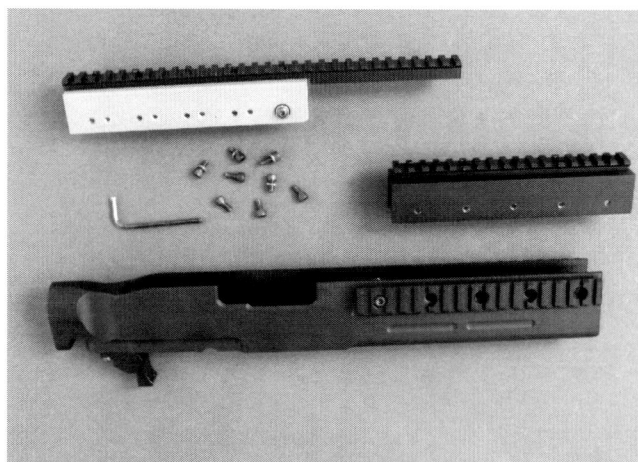

The RB Precision Evolution is a modular stock utilizing a CNC-machined aluminum housing with matching forearm cover. Optional accessories include the long scout rail seen mounted on top cover (above) and standard 7-inch rail that can be readily installed to the sides and bottom of the forearm.

for durability and appearance. The upper half has a matching hand guard made from prefabricated, bottomless, square aluminum tubing and has the same finish as the main body. The rear of the stock is designed to accept numerous varieties of AR-15 stocks and pistol grips. The side and bottom of the forearm have been predrilled and threaded to accept RB's own tactical rails, which are available in 4- and 7-inch lengths. The matching hand guard is also predrilled and tapped for the standard 7- and 11 1/2-inch-long scout rails.

In 2009, RB introduced its Gen-2 version of the Evo stock. As discussed, the Gen-2 is basically the original Gen-1 with the forearm cut off just past the stock assembly screw. A new forearm adapter was fabricated and positioned at the front end of the shortened area, allowing the installation of a tubular free-float-type forearm. This new version is an attempt by RB to upgrade the classic stock design to be more in line with the popular tactical configurations tailored to accept new styles of accessories.

Although the Evo platforms are well made in terms of both materials and craftsmanship, they lack the appeal and authenticity of the AR-15 rifle, even with

the military-style furniture installed. The Evo Gen-1, however, does possess a unique style of its own. Its extra weight and solid overall construction make it the perfect chassis for a serious competition gun.

In the spring of 2011, RB Precision announced on its website that production of the Evolution stocks (Gen-1 and -2) is temporarily suspended. Current information indicates that the Evo stocks production has been put on hold indefinitely and that any info related to these accessories has been removed from the manufacturer's website. There was no specific reason given as to why the Evos have been removed the product line or whether there are any plans to remanufacture these modular stock systems again. If indeed the production of these heavy-duty tactical stocks is over, these stocks are destined to become classic collector pieces in the future, especially for benchrest shooters who prefer these types of stock systems over lighter models.

NORDIC AR-22

The Nordic AR-22 is by far the most attractive of the latest AR-style platforms available today. The AR-22 stock kit comprises the lower (main body), upper body with built-in Picatinny rail, and the fore-

The Nordic AR-22 10/22 tactical chassis is undoubtedly the first of its kind to take virtually any AR-15 furniture and forearms, including one-piece aluminum or two-piece plastic types. The first version (top) of the AR-22 features horizontal grooves on the side, the second (middle) has a border-grooving design, and the third (bottom) version, as used in the Ruger SR22 carbine, is plainer with minor grooving around the magazine chute body.

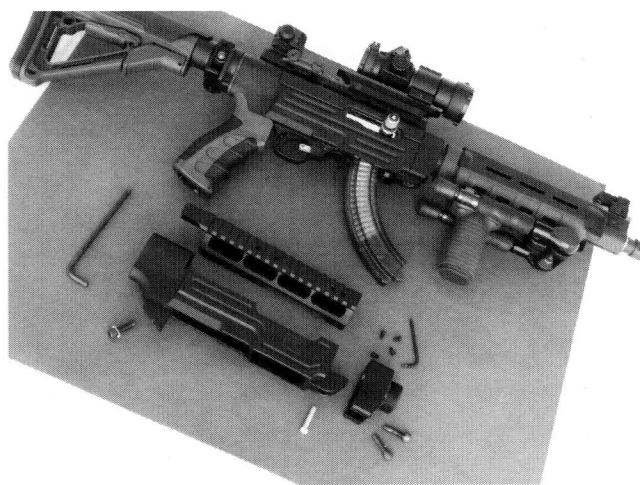

Comparison between the original Nordic AR-22 chassis with horizontal grooving on the side (top gun) compared to the new model with border line grooving (bottom/stripped) showing the three-piece construction. This is the most versatile system of its class to date.

The versatile Nordic AR-22 platform permits the installation of one-piece aluminum free-float forearm (top) or two-piece plastic hand guards (bottom) with the right matching components.

arm adapter. All components of the AR-22 are CNC-machined from billet aluminum with a matte-black-anodized finish. This superb modular chassis system has the perfect geometry of an AR-15 rifle and takes virtually any pistol grip, stock, and forearm used on the main U.S. battle rifle. The AR-22 is the same modular chassis system employed by Ruger in its SR22 tactical carbine.

In 2010, Nordic made several changes to its original design. Much larger screws are now used for the installation of the forearm adaptor. The nylon screws that were supposed to take up play for the receiver inside the platform were found to be unnecessary and have been removed. The original straight-line grooves on both sides of the lower chassis have been replaced by border-style grooving. The company logo is engraved below the ejection port slot. The top Picatinny rail from the early production kit has shallow cross-slots, making it difficult to install an accessory with large assembly screws without leaving a thread mark on the slots. The one flaw that the Nordic design still needs to address is the large, unsightly gap between the upper and lower halves when the two-piece parts are fully assembled. Not utilized in the AR-22 design is the trigger guard gap filler, which Ruger incorporated in its SR22 to improve its overall aesthetics. Despite these minor

flaws, the Nordic chassis system still remains unbeatable in its class.

FABSPORTS FS556

The Canadian-made FS556 is the brainchild of Fabrice Neveau, founder and owner of Fabsports, which is based in Anjou, a suburb of Quebec City. Neveau conceived the FS556 out of frustration, after being unable to import many U.S.-made accessories into Canada because of new restrictions imposed by the U.S. State Department on gun-related parts and accessories. The talented French Canadian utilized his extensive background in machine work and design to create his own chassis system for the 10/22 to compete with its U.S.-made counterparts.

Neveau realized that many kits were already offered for the AR-15 format and opted instead to design a system based on the famous SIG556 assault rifle receiver pattern. (His model was made before the new SIG 522 tactical .22 was introduced.) His FS556 consists of a three-piece design much the same in concept as the Nordic AR-22 but arranged differently. The lower body of the FS is connected to the front and rear of the upper section by two large socket-head screws. When the two halves are fully assembled, the connection is near seamless and very sleek looking. The forearm adapter is connected to the front end of the lower body by five large screws. The adapter comes threaded and does not require a separate barrel nut, making it a much simpler arrangement than the Nordic design, which requires a separate nut during installation of a free-float forearm.

Another good feature of the FS556 design is the fact that the top rail is parallel to the forearm rail, just like on a flattop AR rifle. This allows the installation of a regular gas block to the barrel, thus permitting standard backup iron sights to co-witness (align) properly. In the Nordic AR-22, a higher than normal gas block is required on the barrel to get enough elevation to match the extra-high Picatinny rail on top of the upper half so that the backup iron sights can co-witness.

Despite the many superb attributes of the FS556, it has one serious design drawback: it will not take a drum-type magazine. The SIG556 receiver pattern has a lowered skirt on its magazine chute, and the

The Canadian-made FS556 10/22 chassis is based on the famed SIG 556 assault rifle receiver format. This high-quality CNC-machined aluminum platform was only produced in limited quantities (50). It takes AR-15 furniture and a one-piece aluminum free-float forearm. It is destined to become a collector's item.

FS556 platform faithfully copied this feature. As a result, the extended skirt of the magazine chute prevents full insertion of the drum. All banana-type magazines can be used, as can the Mitchell/MWG 50-shot teardrop-type magazine (with slight beveling required on both sides of the skirt's internal opening to clear the slight tapering at the upper section of the Mitchell/MWG magazine).

In addition to its custom SIG-style chassis, Fabsports also produced matching free-float forearms and slotted endcap. Although, the FS platform was based on the SIG design, it was made to accommodate AR furniture, free-float forearms, and other related accessories.

The FS kit was produced in limited quantities, but if demand warrants its reproduction, Fabsports may consider bringing it back. The company also manufactured a dress-up kit to transform the 10/22 into the famous M1A1 Thompson SMG configuration. Neither the FS556 nor Thompson .22 kit is currently available.

OTHER CHASSIS SYSTEMS

Troy Industries
Another big name in the tactical gun market, Troy Industries, just recently entered the 10/22 custom tactical chassis competition with its ultrasophisticated T-22, in what appears to be a deliberate attempt to give the reigning Nordic AR-22 a run for its money. Troy's kit is available in sport and tactical versions, and appears to exhibit the high quality and workmanship for which Troy is known. The Troy 10/22 chassis comes complete with an A2-style pistol grip, M4 stock, and Troy's own free-float forearm with built-in spring-loaded battle sight, and it is compatible with standard and bull barrels.

Although the new Troy T-22 has very impressive features, the Nordic chassis still holds the edge simply because its overall design features allow the installation of various sizes and brands of free-float forearms, including two-piece plastic hand guards. This advantage provides the hobbyist gunsmith an unlimited capacity to create any weapon configuration he may wish for, from long-range competition rifles down to an ultracompact SBR or PDW. At present, the T-22 only comes in one size format, which is ideally suited for carbine or rifle builds but not for the more desirable compact SBR or PDW sizes for use by Class 2 manufacturers or licensed individuals.

North Eastern Arms
North Eastern Arms (NEA), a dynamic Canadian company, was founded in 2008. A division of a 50-year-old company called North Eastern Aerospace based in the suburbs of Toronto, NEA entered the tactical gun market with the primary goal of designing and manufacturing custom-grade accessories for such popular tactical firearms as the AR-15, VZ-58, CZ 858, M14, H&K, Robinson Arms, Tavor, Swiss Arms (SIG), Remington and Mossberg shotguns, and the Ruger 10/22. In a very short time, NEA has proven itself to be a fast-rising star in the tactical gun market with the introduction of many original, high-quality accessories. Of special note is its CNC-machined aluminum SIG 556 lower and H&K G36 magazine well adapter that permits the use of AR-15 magazines.

NEA has created most of the accessories currently available for the VZ-58/CZ 858 tactical rifles, and some of them are being marketed by other big names in the industry under their own banners. In addition, NEA also created its own version of a chassis for the Ruger 10/22. The chassis is somewhat similar to the

FS556 design but in a much simpler form, with straight-side grooves and without the SIG-style magazine chute skirt. At the time of this writing, NEA was marketing an AR-15 upper of its own design and planning to produce a complete line of tactical AR-15s and a short-barreled version, including one with 7 1/2-inch barrel. A second-generation 10/22 chassis is also being developed and promises to be a winner. For more information about NEA tactical accessories, visit the company's website at www.northeasternarms.com.

CHAPTER 3

Tactical Pistol Grip Selection

The incredible number of aftermarket accessories available for the AR rifle is truly mind-boggling. High-end furniture is definitely what makes custom tactical rifles stand out from basic models sporting standard GI-type gear. For example, buttstocks are available in several shades of tactical colors and design formats, including fixed, retractable, and folding. Prices also differ, ranging from cheap (under $50) to very expensive ($400–$500). The expensive ones are primarily used in military/police sniper-type rifles. These highly specialized stocks normally fea-

ture a cheek rest for height and eye relief for optics and a built-in adjustable leg at the bottom of the butt for long-range precision shooting. For our tactical gun projects, the best choice would be the mid-priced ($100–$150) stocks.

Ideal pistol and forward grips are also mid-priced, as are hand guards. My choices in this category are those manufactured by Magpul Industries and CAA/EMA Tactical. Custom parts made by these companies are well designed and superbly crafted. Their prices are well within reach of the average consumer. These components are also found in many high-end, full-bore tactical rifles marketed by some of the big names in the industry. Why spend $400 for a very expensive accessory when a $100 item can offer the same performance and good looks? Take what you save and invest it in a quality red-dot optic or long-range tactical scope. Remember, a true tactical rifle must not only look good; it must also be extremely accurate.

Among the three major brands of custom chassis used in the tactical gun projects in this book, the Nordic kit requires the least fitting or modification to seat the CAA and Magpul grips properly. These custom grips feature an upsweep or beavertail-type extension at the rear topmost section to enhance ergonomics and comfort over those provided by standard AR-type grips. The RB Evolution stock has an inverted U-shaped cut radius behind the grip seat. Installation of a standard-type AR grip without filling in this cutout section of the chassis will seriously

The tactical gun market is flooded with AR-15-style pistol grips in various styles, sizes, and colors. Models shown are made by Colt, Magpul, EMA Tactical/CAA, Tapco, and Tango Down. Grips with beavertails are cut off to fit the FS556 chassis.

The best grips to use on the AR-22 chassis are the ones with beavertails made by Magpul, CAA, and the DPMS Panther Tactical Grip with palm support (left).

A tactical rifle or carbine is not complete without some sort of forward grip. Forward grips come in different styles, sizes, and colors to match the pistol grip and furniture. Shown are grips made by CAA, Magpul, Tapco, ACE, and SOG Armory.

affect its overall appearance. Luckily, CAA and Magpul have just the right grip to correct this aesthetic flaw.

To fill in the cutout behind the grip mount location, simply cut and fit the top end of the beavertail, following the contour of the radius until the grip seats perfectly in place. A wood rasp/file is ideal when working with plastic materials and is the perfect tool to use here. A regular steel file will barely cut this type of plastic, which means it will take longer to do the job. Reserve the steel file for use as a finisher after rough-cutting the shape of the beavertail with the wood rasp. A shoemaker's rasp is ideal for most synthetics. The smaller the gap that shows between the chassis contour and the modified section of the grip when fully seated, the better the gun will look.

During the cutting and fitting process, the proper way to install the grip to the platform is to first insert the beavertail into the stock cutout, pivot the grip forward to line up for proper seating, and then push it in upward. If the grip does not seat flush underneath the stock, the beavertail needs more trimming. Continue cutting the beavertail in a circular pattern until the grip seats flush in place. Avoid overcutting, especially

in areas where the beavertail needs only minor trimming, so there will be no unsightly gap between the grip and stock after fitting. Always remember, the overall quality of workmanship and the fitting of parts and accessories are what separate the men from the boys. Don't rush; do it right.

The Canadian-made FS556 will also require modification to the beavertail section of the CAA and Magpul grips, as do other brands with this design feature. The cutting operation for the grip is much simpler than that required on the RB Evo stock. In this kit, the base of the beavertail must be cut off and squared at the top, just like on a standard AR pistol grip configuration. If you prefer to use a standard AR-style grip on the FS556, the Tapco Fusion grip is perfect. This economically priced grip possesses excellent ergonomics and comes in variety of tactical colors to match the other furniture. It is patterned after the FN249 SAW machine gun grip, but it is also a near-perfect facsimile of the Swiss SIG 556 assault rifle pistol grip. FS556 authenticity can be further enhanced by using the Fusion grip and an M4-style buttstock mated to an ACE folding stock mechanism.

CHAPTER 4

Hand Guard Selection and Accessories

There are many types of hand guards currently available for AR-type rifles. The Nordic AR-22 kit is designed to accommodate virtually any AR-type hand guard, be it one piece or two piece, aluminum or plastic. The Canadian-made FS556 will only take one-piece, tubular-type hand guards. For plastic hand guards, my primary choice is the Magpul Original Equipment (MOE) hand guard currently offered in all versions: carbine, mid-length, and rifle. To use the MOE guard with the Nordic kit, you will need the complete Delta ring and barrel nut assembly, as well as a round hand guard cap if you are using a carbine hand guard. The Delta ring and barrel nut assembly are installed on the AR-22 forearm adapter as you would normally do on an AR rifle. Use the triangular

The Nordic AR-22 is designed to take not only one-piece free-float forearms, but two-piece AR-15 plastic forearms as well. Magpul recently introduced superb two-piece plastic forearms for AR-type rifles and carbines in various colors, making it possible to create a beautiful two-tone package, such as the one shown on the right.

hand guard cap if you are using a rifle hand guard. This setup will require a .920 outside diameter (OD) bull barrel.

There are two ways to install the two-piece plastic furniture on the AR-22 platform. If you prefer not to modify the heavy barrel, simply enlarge the .750 inside diameter (ID) hole of the hand guard cap to .920 if the outside diameter of the bull barrel is .920 inch. If you choose to use the standard .750 ID hand guard cap, the section of the barrel where the cap is to be positioned needs to be turned .750 OD, starting from the muzzle. This allows the standard-size cap to slide freely to its designated position in the barrel during assembly, thus securing the front end of the two-piece hand guard. Once the cap is locked in against the front end of the two-piece guard, push the gas block against the cap while tightening the set screws during sight realignment. This will tighten and secure the forearm in place against the spring tension of the Delta ring assembly.

Unlike on the AR rifle, the two-piece hand guard assembly is normally disassembled individually to access the gas tube. With the Nordic kit, the two-piece hand guard is kept intact together (fully assembled) with the AR-22 forearm adapter by the Delta ring barrel nut assembly and front endcap. This arrangement is more or less similar in function to a one-piece free-float forearm. To disassemble the plastic forearm group from the gun (assembled as one unit with the forearm adapter), first remove the muzzle device and gas block from the barrel. Disassemble the forearm adapter from the main (lower) chassis and slide the complete forearm assembly forward and off the barrel. The action of the firearm can now be separated from the lower chassis in the usual manner for regular cleaning. Reverse the above sequence for complete reassembly.

For a one-piece free-float-type forearm, my top choice is the Yankee Hill Machine (YHM) brand. This company makes some of the finest aluminum forearms in the industry. They come in many styles and sizes, and their prices are far better than most of the other high-end brands. One feature I like very much with the YHM design is the front endcap. The ventilated cap helps minimize entry of dirt at the front end of the tubing and also enhances the overall aesthetic of the hand guard and the gun in general.

If you are building a tactical gun with two-tone accents to complement the buttstock and pistol grips, the hand guard (be it one or two piece) can also be accentuated to match the furniture. When using a plastic hand guard, such as the Magpul MOE, you would obviously need two different-colored guards to match the color combination of the grip and stock. You simply mix and match the upper and lower sections of the guards. Check which color combination looks best with the gun among the buttstock, pistol grips, and forearm. Look at it from close range and from a few feet away, and from all angles, to determine if the color combination is what you want. If your sixth sense tells you something is not right, find out what is missing and reevaluate the setup. You may have to alternate the accessories until you find just the right combination for your creation. Firearm accessories are like jigsaw puzzles—they must coordinate perfectly with the other components to look good and function harmoniously.

With the one-piece aluminum hand guard, you can use the ladder-type rail protectors available in popular tactical colors to match the furniture. If your primary color is green, then use green protectors. If you want to add a forward grip, green is the ideal match. Then you must decide whether you would also like to have a two-tone accent on it. Use black skateboard tape for this purpose. The more you practice this type of gun art, the better you become, and soon you will be even more creative when coordinating the colors on your project guns.

There is so much to explore in this amazing world of tactical weaponry, so be creative in every aspect of assembling your weapon. Imagine the endless array of tantalizing parts and accessories available out there and the things that you can do with them.

The tactical gun market is currently flooded with one-piece, aluminum, free-float forearms for the AR-15 rifle and its derivatives. I found the Yankee Hill Machine (YHM) forearms to be the best value in terms of style, quality of workmanship, and price. Shown here are just a few of their extensive lineup of free-float forearms in various lengths. Tubing is available in customizable format (plain) or with built-in tactical rails.

The YHM customizable, aluminum, free-float forearm does not have a built-in rail like other models. It was tailored in such a way that a separate rail can be added if required in any of the four sides of the tubing, which has predrilled holes (threaded) ready for mounting the size of rail required (comes in various lengths).

CHAPTER 5

Collapsible- and Folding-Stock Selections

The M4-style collapsible buttstock is undoubtedly the most popular tactical stock on the market today. It is found on military and police firearms, tactical shotguns, and sporting rifles. The vintage CAR-15 stocks of the 1960s, from which the new M4 stocks got their inspiration, normally have three adjustment settings, while the modernized versions have six. A rifle with an M4 stock is about 4 inches shorter when the stock is fully collapsed.

The good news is that you don't need to spend a fortune buying the most expensive collapsible stock for the tactical 10/22. If you have an AR already equipped with a good buttstock, you can use it on your tactical 10/22 to get the same feel as on your full-bore battle rifle. My criteria for a tactical stock are that it be modern, well made, and sleek, and not overly decorated. Above all, it must be comfortable and user friendly. The price tag is also a consideration.

My top choice in this category is the CAA CBS six-position collapsible stock. This Israeli-made buttstock features side battery storage and Picatinny rail to accept the CAA Adjustable Cheekpiece (ACP). The CAA CBS stock is available in black, green, and tan.

I also highly recommend three Magpul stocks for the tactical 10/22: Magpul Original Equipment (MOE), Compact/Type Restricted (CTR), and Adaptable Carbine Storage (ACS). MOE is the company's economically priced version. It provides quality,

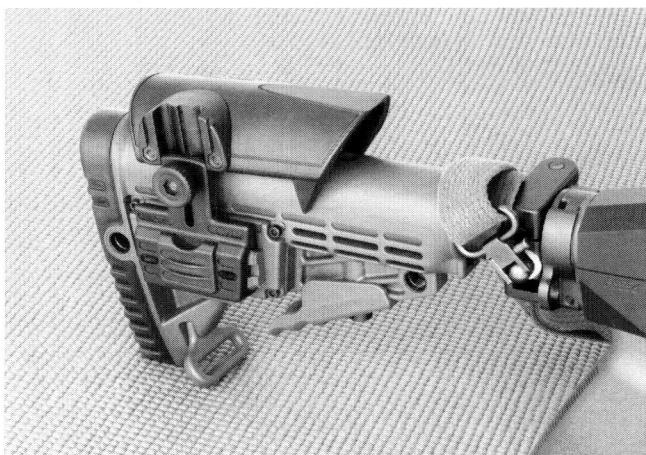

The versatile Israeli-made CAA/EMA Tactical modular buttstock features a side-mounted rail, which allows the installation of such additional accessories as the adjustable cheekpiece shown in the raised position.

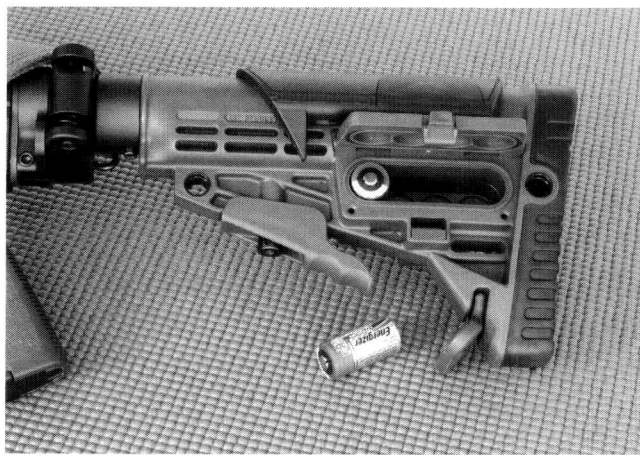

The opposite side of the CAA buttstock, showing the battery-storage compartment for four CR123 batteries. The lid is shown hinged open. Note the soft, rubberized butt pad designed to absorb recoil.

Another excellent accessory from CAA/EMA Tactical is the M4 stock saddle available in standard or rubberized versions. This accessory provides comfortable cheek support and battery storage on both sides. Note the Midwest Industries (MI) AR sling adapters shown below the stock. The CAA One-Point Sling (OPS) is connected to a CAA sling adapter (on gun).

The CAA/EMA Tactical FAL carbine stock cheekpiece provides correct eye-level shooting comfort when the gun is equipped with high-mounted scope, such as this AR-22 with carry handle sight.

Tactical buttstocks made of plastic (CAA, Magpul, Lewis Machine and Tool, or LMT) or metal (ACE) can be accessorized to meet specific requirements. Some of the accessories shown are adjustable and fixed cheekpieces. The CAA stock saddle mated to a standard M4 stock (background) doubles as battery storage.

The Magpul snap-on stock cheek riser is a perfect combination for CTR/MOE buttstocks. It is available in 1/4-, 1/2-, and 3/4-inch increments. It provides added support for correct eye alignment with backup iron sights or optics. The 3/4-inch cheek riser is shown installed to the CTR, but the stock shown is in the folded position. The high cheek riser is ideal for long-range scope installation.

durability, and the ergonomic benefits of the upgraded CTR. The CTR differs from the MOE design in that it incorporates a new friction-locking mechanism that eliminates or reduces movement between the buffer tube and buttstock for solid feel when the weapon is fired from the shoulder. The CTR accepts a variety of ambi-sling attachment options. The latest mid-priced collapsible stock from Magpul is the ACS. This beautifully designed stock gives the operator extra storage options and a secure, comfortable cheek weld, complemented by advanced dual-locking levers for maximum stock rigidity. The MOE, CTR, and ACS stocks

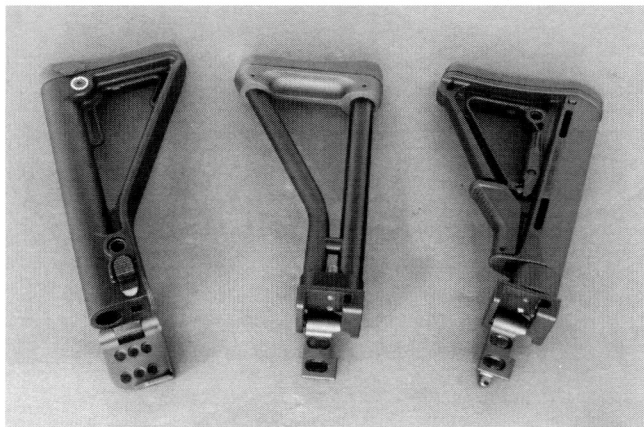

Adding a folding-stock option to your .22 tactical hardware not only makes a long gun more compact for storage and transport, it also increases aesthetic and overall value. These rugged folding stocks are made by AGP (plastic; left), ACE FN-FAL paratrooper-style FX-FS (middle), and Magpul CTR with ACE folding-stock mechanism (you have to build this yourself).

are available in black, dark earth, olive drab, and foliage green.

The advantage of a side-folding stock is obvious, especially on a long-barreled rifle (such as a sniper or competition model): compactness, which makes the rifle easier to transport and deploy. I also appreciate the edge offered by a folding stock, along with the versatility of the M4 collapsible buttstock. With these two superlative components, you have the best of both worlds.

To build this unique stock system, you will need the following ACE parts:

- CAR stock (PN)—accommodates M4 stock extension tube.
- Receiver block (ARRB)—assembled to the receiver; base to folding stock mechanism.
- Folding stock mechanism (FSM).

If you prefer a one-piece design rather than the three-piece stock system, you can order the ACE folding mechanism with integrated AR-15 stock interface (FSM-AR). This part connects directly to the back of the chassis, and the rear-threaded extension allows direct installation of the M4 stock extension tube. This is a single setup dedicated solely for

AR rifle buttstock installation. The three-piece design is a more versatile arrangement for hobbyist gunsmiths who like to experiment with various stock installations geared toward their specific design requirements. This setup will accommodate the collapsible M4-type stock, A2 fixed buttstock, and the FAL-style skeleton metal stock.

READY-MADE ACE FOLDING STOCK SETS

ACE, Ltd., recently introduced complete ready-to-install folding stock sets for the AR and SIG 556 tactical rifles for those who do not wish to build their folding-stock system from the ground up. Four models are offered, each having a different buttstock profile:

- ARRB-FX-FS (with FAL side folder but in a much more beefed-up format)
- ARRB-M4M-FS (with M4-style collapsible buttstock)
- ARRB-M4S-FS (with modular SOCOM stock)
- ARRB-UL-M-FS (with ultralight modular stock)

These fully assembled stock sets are of the highest quality. They are well-designed side-folding stock systems specially tailored for wide variety of tactical applications. They are a bit pricey, but you certainly get your money's worth and you save time by not having to put it together yourself.

Another nice folding stock just introduced for unconventional AR-15s with their recoil spring not dependent on the stock extension tube is the AGP side-folding stock. This folder is ideally suited for AR rifles with dedicated .22-caliber conversion kits. It is also perfect for 10/22s utilizing the Evo, AR-22, and FS556 custom chassis. Made of glass-filled nylon material, this stock is light, strong, and stylish. A matching rubber butt pad is available directly from the manufacturer as an option. The AGP stock comes with a stock adapter and an alignment bushing. The AGP adapter is similar to that of the ACE ARRB design and is also assembled directly to the rear of the chassis.

To achieve the necessary tightening effect between the connection of the adapter and the stock base when assembled to the chassis, the top end of the adapter

Close-up of the ACE ultrastrong folding-stock mechanism combined with Magpul CTR buttstock, shown in the folded position. This hybrid stock system is the best of both worlds, providing collapsibility and folding capability for the stock and thus maximum compactness on a long-barreled rifle.

Close-up of the AGP folding stock, showing the four assembly screws that connect the stock base to the rear of the receiver. Note the large push-button locking mechanism and built-in sling attachment. The stock has no locking capability in the folded position. It relies on the friction of the pivot pin to hold it in place, which is less than ideal.

must be positioned slightly lower than the edge of the assembly hole. The cutout for the alignment bushing in the adapter must be at the 6:00 position when the stock is assembled to the adapter. There are four large socket-head screws that secure the stock to the adapter. In addition, the stock has two built-in ambidextrous QD-sling swivel attachment holes, one behind the push-button latch and the other in the upper section of the butt. At present, the AGP stock has no provision for adjustable cheekpiece installation.

NOTE WHEN INSTALLING THE AGP STOCK TO AR-22 CHASSIS

The hole at the back of the AR-22 chassis where the alignment sleeve for the AGP stock would seat is drilled deep and needs a 3/8-inch-diameter by 2-inch-long utility spring (available from most hardware stores) to push it back in order to lock the sleeve firmly against the base of the stock during assembly.

The installation of the AGP folding stock to the Nordic AR-22 chassis will require a 3/8 x 2-inch spring to lock the alignment bushing against the stock prior to tightening the four assembly screws as shown.

CHAPTER 6

Mix and Match: Two-Tone Furniture Combinations and Custom Accents

The colorful array of tactical firearms accessories today is truly refreshing. Buttstocks and pistol grips made of plastic materials are available in a variety of tactical colors. I consider the two high-quality brands of plastic furniture (buttstocks and pistol grips) made by Magpul and CAA the best choices among the many competing brands primarily because of their outstanding design features, excellent workmanship, precision fitting, and unbeatable prices. CAA buttstocks and pistol grips are available in black, green, and tan, while Magpul's similar accessories come in black, dark earth, foliage green, and

olive drab. These two brands differ in styling but are equally sophisticated and ultramodern.

The CAA Universal Pistol Grip (UPG16) for AR rifles features multiple finger grooves and back inserts, as well as an internal grip compartment. Similarly, Magpul has the Mission Adaptable (MIAD) modular grip system. The MIAD also features interchangeable grip inserts and two types of grip core that slide inside the grip compartment. The individual cores can store spare batteries or an M16 bolt and firing pin.

Stunning two-tone grips can be made from these

Creating a beautiful two-tone color combination with an AR pistol grip can now be easily accomplished using grips that have interchangeable back and front inserts, such as the CAA UPG16 (shown) and Magpul MIAD modular grip system.

Color coordination is one of the major benefits of today's ultramodern tactical weapon furniture. EMA Tactical/CAA designs its furniture pieces in such a way that they can accommodate various accessories to enhance their versatility and effectiveness, with the added bonus of two-tone color combinations, as shown with these stocks and pistol grips.

two brands simply by purchasing several grips with different colors. The best two-tone color combinations are black and green, tan and black, dark earth and black, olive drab and black, and foliage green and black. Green and tan or dark earth and olive drab and tan simply don't quite match, no matter what the finish of the gun is. With handguns, you may get away with it but definitely not with a rifle. These two-tone combinations must be used in conjunction with the buttstock and forearm color combinations as well.

With the Magpul CTR and MOE buttstocks, the optional stock cheek riser accessory is a perfect accent. This cheek riser is available only in black but will go nicely with any of the company's tactical colors. The Magpul cheek riser is available in 1/4-, 1/2-, and 3/4-inch increments. When opting for a tan CAA buttstock, use CAA's optional adjustable cheekpiece in black as an accent. Combine this with a tan pistol grip with black inserts. If you are using a one-piece, free-float forearm that has a black-anodized finish, install a tan forward grip and tan ladder-type rail protectors. If the forearm is a two-piece plastic version, use a tan-colored forearm at the bottom and a black one for the other top half, or vice versa—the choice is yours. You would have to buy two pair of two-piece hand guards of different colors to be able to do this color combination. Another option is to paint the other half if you cannot get a colored forearm for the size you need. If you want a two-tone look with a forward grip, use black skateboard tape (available from most sporting goods stores that sell or repair skateboards). If the gun is painted olive drab, foliage green, or another shade of green, the buttstock and grip should coordinate with black inserts.

Having various styles and colors of buttstocks, pistol grips, and forwards grips on hand is a tremendous advantage when using one of these custom AR-22 platforms. When you get bored with one setup, simply change the parts and, presto, you have a new gun to show off and enjoy. With so many accessories coming out for both the 10/22 and AR-15 rifles, you can customize and modernize your tactical 10/22 whenever you want. This is one advantage those competing factory tactical clones can never match.

Forward grips can be tailored to match the color combination of the pistol grip and plastic buttstocks using O rubber rings and skateboard tape. Magpul's AFG (installed on the gun) has removable inserts, which allow you to simply interchange the inserts to match different-colored grips.

Pistol grips can now be matched with other accessories, such as backup iron sights, as in the case of the Magpul grip (installed on gun) and MBUS rear sights, which are made of polymer materials. Other grips shown are the CAA UPG16 (with beavertail removed) in two-tone with target stand (center) and Tango Down BG16 grip (right).

A prime example of a superb two-tone combination (dark earth and black) using Magpul CTR stock, MOE pistol grip, and MBUS sights. The MOE pistol grip utilizes a precut, stick-on, black-rubberized decal, complemented by an MOE/CTR stock cheek riser.

Magpul's Mission Adaptable (MIAD) modular grip system for the AR-15 features interchangeable inserts. They are available in black, dark earth, and foliage green. You can create a beautiful two-tone grip by combining these colors to match the buttstock and its accents. Seen here are black and dark earth combinations.

Another superb two-tone color combination (tan and black) using CAA/EMA Tactical furniture. The stock has a black adjustable cheekpiece, the pistol grip has black front and rear inserts, and the matching forward grip (shortened VGI grip) has skateboard tape on both sides and between the finger grooves.

Magpul's Angled Foregrip (AFG and AFG-2) also features interchangeable inserts just like its MIAD pistol grips and can be made in two-tone format by combining different color grips. Pictured is the black and dark earth combination to match the pistol grips and buttstock accents.

Close-up of the CAA green UPG16 pistol grip with black inserts, complemented by CAA's SVG grip with rubber O-rings assembled to the circular grooves to create a beautiful two-tone accent that matches the pistol grip.

CHAPTER 7

Backup Iron Sights, Optics, Gas Blocks, and Mounting Rails

Backup iron sights and optics are two of the most important elements for a true high-performance tactical rifle. The gun can look fantastic, and it can be amazingly reliable, but if it's not hitting your target, you'd better hang it on the wall for display purposes.

The 21st century has ushered in many advancements in technology, including innovations for firearms and related accessories. Optics and iron sights particularly have enjoyed an amazing leap beyond their vintage counterparts. Battle optics, such as ACOGs and EOTech sights in particular, have become essential for modern foot soldiers, who rely on them to be able to hit their targets quickly and accurately at various distances, ranging from up close to several hundred meters away. Long-range optics have proven their worth in the hands of professionally trained marksmen in actual combat with .50-caliber Barrett M82/M82A1 sniper rifles and other exotic hardware designed to hit targets at distances previously thought impossible.

Backup iron sights are the perfect companion for battle optics on standard rifles (M16/AR-15), carbines

Backup iron sights are among the hottest items in the tactical gun market today. Shown are just a few of the many variations of flip-up sights available for flattop AR rifles: the Magpul Gen-1 MBUS (FDE plastic), YHM (steel), and Brownells (aluminum).

Close-up of the Magpul Gen-1 MBUS forearm front sight (shown here unassembled) in flat dark earth color. This lower front sight will not work with Nordic AR-22 unless the forearm rail is the same height as the platform's upper rail. YHM has developed a gas block riser accessory (YHM-230), seen mated with the DPMS .936 four-rail gas block, which elevates the riser to the same plane as the forearm rail. The taller gas block front sight in this setup will now co-witness with the rear sight using the Nordic AR-22 platform.

(M4), short-barreled rifles (SBR), and personal-defense weapons (PDW). The advantage of a backup iron sight system is that it acts as an emergency sight in the event the optic becomes inoperable, which usually happens when the battery is dead or the optic itself has been damaged beyond repair.

The introduction of the universal rail system in the tactical gun market is truly a remarkable achievement in small-arms innovation. It has benefited not only modern military- and police-type weapons, but also civilian sporting guns, hunting rifles, shotguns, and even handguns. Installation of backup iron sights requires a weapon rail as a mounting platform. Most tactical firearms have a built-in rail at the top of the receiver to accommodate this accessory in conjunction with optics. Free-float forearms and tactical gas blocks for AR-15 rifles are mostly offered with matching rails to accommodate the front sight.

When you are building a tactical 10/22, it is best to install the front sight on a gas block rather than on the forearm itself, as on the AR rifle, simply because the barrel of the 10/22 is not always perfectly concentric with the center of the forearm when using the AR-22 stock chassis. On the SR22, Ruger incorporated a barrel support block to center the barrel at the front end of the forearm. This support block only works with Ruger and Nordic forearms. The FS556 excels in this critical area because the design leaves enough room inside the chassis for the receiver to be adjusted in order for the barrel to line up perfectly

This versatile Canadian-made FS556 10/22 is equipped with a UTG tactical rear sight, complemented by YHM scope riser and Bushnell holographic sight, a derivative of EOTech red-dot scopes. This gun features a CAA G16 finger-groove grip, complemented by CAA's new CQB magazine grip.

Close-up of the Pro-Mag AR-15 flip-up front sight and the versatile Double Star Carlson AR-15 adjustable muzzle brake. The Pro-Mag sight is quite attractive, but its push-button lock is flimsy. The front sight post thread is not compatible with U.S.-made parts. Note the YHM extended rail riser (YHM-234) added to the bottom rail of the forearm to accommodate the TLR light.

This custom FS556 SBR works well and looks good with muzzle devices from Yankee Hill Machine, Fabsports, and Tactical Innovations. Note the YHM customizable free-float forearm with matching forearm endcap. The tactical light is from NcStar. Th (LMT forearm-mounted tactical front sight sits perfectly on the MI high-rise gas block. The quick-detach sling swivel mount seen behind the sight and magazine grip is from EMA Tactical.

straight with the free-float forearm. In the AR-22, you can easily see this problem of the barrel not lining up straight with the axis of the forearm by looking at the front end of the free-float tube when the gun is fully assembled. The gun is also more accurate

This superb tactical 10/22 rifle is dressed with an AR-22 Gen-2 chassis complemented by Magpul furniture. It looks great with the Rock River Arms (RRA) tactical carry handle with A2 rear sight. It is best to use a low-mounted, compact, red-dot scope, such as the Tasco Propoint 1x32mm rifle scope on the carry handle rail for attaining correct eye level without a stock cheek riser.

if the front sight is assembled directly to the barrel itself—in this case, with the aid of the gas block.

The market is currently flooded with superlative backup iron sights, and their prices vary according to name brand, materials, and quality of workmanship. Preferably, the ideal backup iron sight combinations for building 10/22 tactical guns are those equipped with windage and elevation adjustments for the rear sight. Since these sights are designed primarily for AR-type rifles, the rail on top of the 10/22 receiver may not be perfectly compatible with the height of the forearm rail or gas block rail-mounted on the barrel. The extra elevation of the rear sight and the front sight normally corrects the problem when the backup iron sights do not co-witness properly.

Among the three chassis systems (RB Evo, Nordic AR-22, and FS556), it is the Evo stock kit that definitely requires a rear sight with elevation and windage adjustment. Although a Hahn Precision flattop rail is mounted on top of the 10/22 receiver, it is a tad low to perfectly co-witness with the front sight, whether it is installed to the hand guard rail or gas block. The use of an A2-style fully adjustable rear sight is definitely a must for this gun setup, as it may require elevating the sight to its maximum height in

Variation of an AR-22 match rifle equipped with YHM's solid A2 rear sight that sports an EOTech riser. The EOTech 512 red-dot scope is seen below the gun. This beautiful tactical rifle is dressed in CAA two-tone (tan/black) furniture, complete with ACE's rugged folding-stock mechanism. Note the target-shooting stand (UPGS) mated to the CAA UPG16 pistol grip.

This colorful two-tone "competition ready" Evolution 10/22 is equipped with YHM solid A2 rear sight (YHM-643K) mounted on the Hahn Precision low-profile riser—an AR-15 flattop riser accessory. The standard peep sight has been replaced with DPMS hooded aperture rear sight for long-range application. Note the CAA VG1 forward grip, complemented by skateboard tape on the side and front.

Shown are just a few of the vast array of rail risers used on flattop AR-15s and other tactical-type rifles and shotguns. These accessories come in different styles, lengths, heights, materials, and prices. These risers are from Hahn Precision, CAA Tactical, and Brownells.

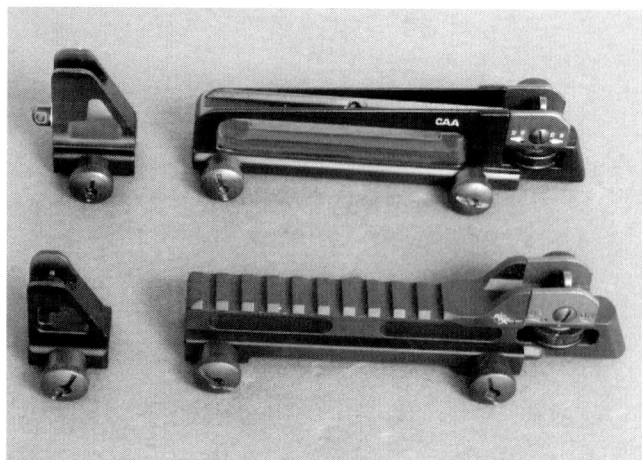

For those who may still prefer the classic AR-15 configuration, the economically priced CAA Carry Handle (CH) sight with matching UTG gas block–mounted front sight is the ideal combination (top). RRA made a refined, expensive, and more sophisticated version of the CH called the Tactical Carry Handle with A2-style rear sight (bottom). The RRA rear sight is a perfect match for the LMT forearm-mounted tactical front sight assembly.

order to compensate for the slightly lower receiver rail to achieve optimum accuracy.

I have selected several brands in this category, based on style, quality, and price. The most economically priced backup rear sights tested are the CAA Carry Handle (CH) and UTG/NC Star MARDRS. The CH and UTG/MARDRS are Chinese-made copies of U.S.-designed accessories. These tactical sight imitations do not exhibit the high degree of quality and workmanship of their U.S. counterparts, but they do offer good value for the money.

If you want American-made backup iron sights but must watch your budget carefully, the DPMS detachable rear and front sight combo is unbeatable. The DPMS rear sight is machined aircraft-grade aluminum with a mil-spec A2-style sight that is fully adjustable for windage and elevation. It uses two thumbscrews to secure the sight to Picatinny or Weaver-type rails. The matching front sight is machined from steel and is patterned after the classic fixed front sight styling but in tactical, detachable format.

In the mid-price range, the machined aluminum and steel backup iron sights from YHM are definitely the top choice of both tactical gun builders and high-end AR rifle manufacturers. YHM is famous for its quality, workmanship, ruggedness, styling, and price. A superb tactical backup sight I found to be the per-

This classically styled tactical AR-22 rifle utilizes the CAA Carry Handle sight complete with UTG Leapers SWAT tactical mil-dot scope mounted on a matching CAA Carry Handle Picatinny scope mount. To achieve the correct eye level while aiming with this high-mounted optic, your stock will need a cheek riser.

The CAA Carry Handle Picatinny scope mount, complete with Tasco red-dot scope removed from the Ruger SR22 Carbine. Complementing the Hogue rubber pistol grip is the SOG Armory graphite vertical grip.

This unique sighting setup combines the CAA Carry Handle sight and MI A2 Adjustable Cantilever scope mount. The cantilever places the optic on just the right level as that on a flattop AR, thus eliminating the need for a cheek riser for the stock.

fect complement for a world-class 10/22 tactical setup is the RRA tactical carry handle. Matching this high-end accessory is the LMT forearm-mounted tactical front sight assembly. The sight is similar to the DPMS detachable front sight, but it is shorter and made of aluminum instead of steel. You may also install the LMT part directly to a gas block as long as it is the high-rise type, such as the YHM-9388 (for bull barrel) and Midwest Industries (MI) tactical high-rise gas block for barrels turned to .750 inch OD.

This beautiful tactical AR-22 rifle is equipped with DPMS mil-spec A2-style rear sight complemented by a YHM scope riser (YHM-227A). The red-dot scope is the Bushnell Trophy. Note the two-tone CAA UPG16 grip and Magpul two-piece plastic forearm. The matching forward grip is the short TAPCO Intrafuse vertical grip.

These modern tactical rear sights are true works of art. They are designed to deliver the utmost in accuracy and precision from today's battle rifles no matter what caliber they are. Shown are just a few of the many fine examples of fixed tactical backup iron sights made by DPMS, YHM, and UTG (a cheap copy of the LMT tactical adjustable rear sight).

Close-up showing the LMT forearm tactical front sight (seen detached here to show detail of the gas block rail). The raised gas block places the rail parallel to the forearm rail allowing the lower tactical sight to co-witness with the flattop AR rear sight. This arrangement also works well with the FS556 and Evolution tactical rifle setups.

Shooters who still prefer to use iron sights rather than optics can enhance the sight picture by simply replacing the regular steel front post with one that has a fiber-optic insert. I can recommend two excellent replacement fiber-optic front sight posts for the AR-15: the top-of-the-line Mounting Solutions Sight-Link and the more economically priced Hi-Viz version. The latest entry in fiber-optic sight innovation is a battery-powered unit made by Tru-Glo. This high-tech AR-15 tactical front sight assembly features a push-button switch that activates the battery-operated fiber-optic illuminator to maximize the brightness of the fiber optic in extreme low-light conditions. Another replacement front sight post worthy of consideration is the XS Sight System. This economically priced replacement front sight post features a double-face white outline and is available in two blade thicknesses. These replacement custom sights posts are available from Brownells.

For the Nordic kit, you may use any standard rear backup iron sights of your own liking. The high-rail design of the Nordic kit upper half requires the taller gas block–mounted front sight, even when it is installed to the forearm itself, in order to co-witness with the rear sight. If you opt to install the front sight to the AR gas block, you must use a high-rise gas-block profile.

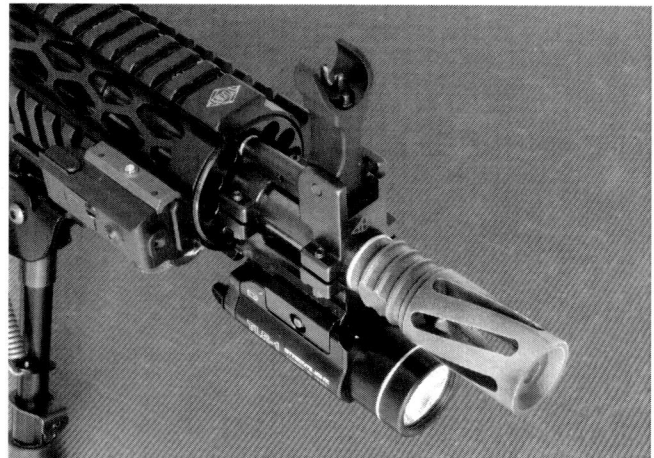

Close-up view of the Hi-Viz fiber-optic sight installed on the YHM hooded front-flip-sight tower. This optional accessory maximizes the effectiveness of the backup iron sights by increasing the clarity of the sight picture in both normal and low-light conditions. Note the Canadian-made (NEA) AR-15/M16 hybrid brake featuring unique spiral slots for muzzle control.

If you are using a heavy barrel, you will need the YHM railed bull-barrel gas block (YHM-9388). The top rail of these gas blocks will be parallel to the height of the forearm rail when installed on the barrel. The YHM-9388 has a bore diameter of .936 inch, whereas the 10/22 bull barrel block averages .920 inch OD, not quite a perfect match but fairly close with just an .0008 gap. Installation of the YHM gas block on the 10/22 bull barrel requires a sleeve to fill in the gap.

A simple but very effective solution to this problem is the use of scrap aluminum from pop cans. Simply cut both ends off the can with scissors and then slice the can vertically to make a sheet. Cut the sheet to measure 4 1/2 inches long by 1 1/2 inches wide. Roll the narrow portion of the aluminum sheet around the barrel and hold it tight where the gas block is to be mounted. Insert the gas block through the rolled aluminum sleeve and slowly wiggle it around until the entire length of the gas block is fully seated in its desired position. If the part is too tight to assemble, you may trim the length of aluminum sheeting (gradually) until the gas block can be forced in with reasonable tightness over the sleeve.

If you prefer to lathe-turn the front section of the barrel to .750 inch OD for standard gas block installation, the MI Picatinny rail gas block is the perfect part to use. The area to be turned depends on the length of the barrel and the type of forearm you are going to use. The stop point for the turned section of the barrel must be in line with the front

One of the latest accessories for mounting optics on a flattop AR-15 is the Leupold Mk-1 integral mounting system, seen here installed on the AR-22 complete with Bushnell Trophy red-dot scope. The height of the mount is perfect when used in conjunction with the YHM A2-style detachable rear sight or similar backup iron sight system.

end of the forearm, where the gas block will be positioned when installed.

The MI gas block is precision-machined from a tough 4140 billet of steel. It is designed to outlast conventional aluminum gas blocks many times over from erosion caused by gas pressure on AR rifles. In the 10/22, a lighter aluminum material would be preferable since the gun itself works on a simple blowback principle. However, only MI makes the high-rise type with .750 ID gas blocks at the moment.

For the RB Evolution stock and FS556 chassis, the standard-height gas block is compatible. However, the high-rise gas block required for the Nordic kit can also be employed with the RB and FS setups if preferred. In this case, instead of using the regular gas block–mounted front sight, you use the shorter forearm-mounted front sights in order for the sights to co-witness. The standard .750 and .936 AR-15 gas blocks are more common, so there are plenty of styles from which to choose.

For the 10/22 setup, ideally you need a top rail for mounting the front sight. There are two affordable, high-quality brands for this option: the DPMS and Brownells modular gas blocks. The DPMS features an integral four-rail design and is available in standard .750 and .936 sizes. The more versatile Brownells version is a modular type with a fixed top rail and optional three-rail system that can be readily installed on both sides and the bottom of the block to meet specific tactical requirements. The Brownells accessory is only available in .750 ID format. This part was precision-machined from 6061 billet aluminum for extra strength and hard anodized to mil- **Note:** when installing the optional side or bottom rail for this gas block, remember not to overtighten the screws, as it may strip the threads.)

Tactical mounting rails are now installed on almost every modern military-type weapon. The versatility of this accessory is simply amazing. It can transform an ordinary firearm into a high-performance combat arm simply by permitting the installation of such tactical accessories as optics, lasers, forward grips, bipods, and so on. The market has an endless array of tactical rails available for virtually every type of firearm you can imagine. I again selected the YHM brand as tops for this accessory because

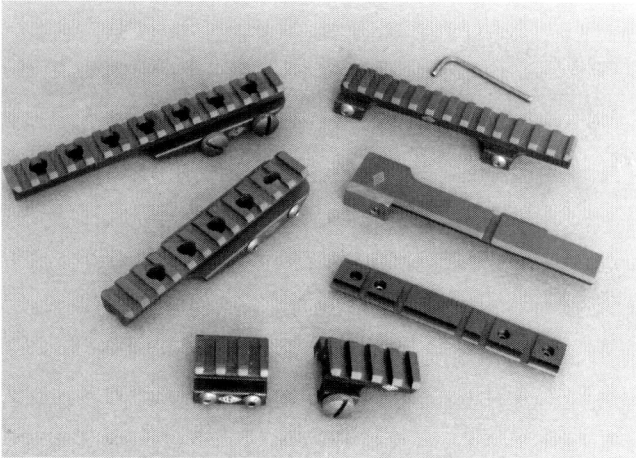

YHM was a frontrunner in tactical accessories innovation, specializing in high-tech aluminum forearms and rail risers. Shown are an assortment of its superlative scope riser, EOTech riser, rail extension, and gas block riser. For quality of workmanship, design, and price, YHM products are unbeatable.

of its quality, selection, and prices. YHM's versatile rails are precision-machined from aircraft-grade aluminum and hard-coated to military specs. Rails used in the guns in this book include the following:

- Scope riser (YHM-227A). This particular accessory is perfect for varmint rifles. Adaptable for all types of scopes, red dots, and other optics, it raises the optic 1/2 inch and mounts to any Picatinny rail.

- EOTech riser (YHM-220). This special mount elevates the EOTech by .300 inch and allows the operator to use his iron sights in the lower third of the optic window.

- Rail extension. This accessory is available in 5-inch format (YHM-9474) and 6-inch format (YHM-9473). This part is ideal when the gun has run out of rail space, especially for short forearms. In the RB Evo UMR gun project, the 6-inch rail extension was used for the gas block to allow mounting a Harris bipod adapter (YHM-638) and a laser or tactical light.

- Gas-block riser (YHM-230). This unique

Variation of an AR-22 match rifle equipped with YHM solid A2 rear sight that sports an EOTech riser. The EOTech 512 red-dot scope is seen below the gun. This beautiful tactical rifle is dressed in CAA two-tone (tan and black) furniture, complete with ACE's rugged folding-stock mechanism. Note the target-shooting stand (UPGS) mated to the CAA UPG16 pistol grip.

Tactical rails can either do wonders for your custom gun or ruin it if you do not know how to use them properly. The PSG22 uses multiple rails to accommodate optic, light, laser, forward grip, and bipod, using the Nordic AR-22 Gen-2 chassis and Magpul furniture in a two-tone (olive drab and black) combination. This is what a world-class tactical 10/22 is all about!

accessory raises the height of a standard gas block to the same plane as the AR flattop or forearm rail. This part has one cross-slot and will only accommodate the YHM Same Plane Front Sight. If this part is to be used in the Nordic kit, the gas block–mounted front flip

Representative AR-15 gas blocks include Brownells four-rail modular type (left), RRA gas block sight base (front center), and DPMS .936 ID four-rail type with YHM gas block riser installed (right). In the background is the YHM hooded, front-flip-sight tower with bottom rail (YHM-9835-H). **Note:** The .936 gas block will require a pop can wrap-around shim to tighten the gap with .920 OD barrels as used on heavy-barrel 10/22s.

Close-up showing the MI high-rise Picatinny rail gas block mated to Brownells AR-15 tactical flip-up gas block front sight. Note the Tactical Innovations FS22 fake suppressor and SIG mini laser-mounted on the side of the YHM diamond free-float forearm.

Two brands of high-rise gas blocks are available today: MI (front center) and YHM (right). The MI gas block has a .750 ID and is made of steel, while the YHM version has a .936 ID and is made of aluminum. The DPMS four-rail heavy barrel gas block will require the YHM gas-block riser to have the same height as the MI and YHM gas blocks as shown (background).

Close-up of the Brownells modular gas block (.750 ID). This versatile AR-15 gas block allows installation of three additional rails (both sides and bottom) to accommodate a multitude of tactical accessories, such as a laser, light, and bipod. Note the Brownells tactical flip-up front sight with Sight-Link fiber-optic sight, complemented by a Troy Medieval muzzle brake.

sight (YHM-9584 or Brownells M4 tactical flip sight) must be used so that it will co-witness with the rear backup iron sights.

• Brownells AR-15 flattop riser. This superb accessory is perfectly suited for the installation of a full-size scope to be used in conjunction with folding rear sights, such as the YHM flip rear sight or

Close-up view of the YHM hooded front-flip-sight tower with bottom rail. This versatile steel gas block features a built-in folding front sight and bottom rail, seen with SIG SAUER compact pistol laser. Note the high-performance DPMS Meculek AR-15 muzzle brake and the TLR light mounted on the side rail of YHM Specter Diamond Series free-float forearm.

If you are building a custom AR-22 or SR22 with the gas block installed on the barrel, use the MI high-rise Picatinny rail gas block on a .750 OD barrel (as shown). For the backup iron sights to co-witness, the standard (taller) gas-block front sights—such as the DPMS/UTG nonfolding, detachable front sight (installed) or the Brownells AR-15 tactical gas block folding front sight (resting on hand-guard)—must be used.

Magpul MBUS polymer folding sights. For fixed-type sights, such as the YHM A2 rear sight, the primary mounting rail where the sight seats will need a half-inch scope riser, such as the YHM-227A. The Brownells flattop riser is then mounted on top of the YHM riser to get the perfect clearance height and allow the rear of the full-size scope to rest over the rear sight; it can be moved to a more comfortable position for the operator.

Naturally, there are much higher scope risers that can do the job of both the YHM and Brownells rail combination. However, since these two accessories are also usable in other tactical combinations, why spend more for additional rails when you can combine these existing accessories for these applications? I call this technique "rail stacking." The system involves stacking one rail on top of the other to increase the height of the mounting platform to meet certain requirements.

This is what makes these tactical rails so versatile. All it takes is a little imagination to maximize their applications in configuring multiple gun formats, all without spending more money for additional parts that aren't needed.

The RRA tactical carry handle A2 rear sight has a built-in rail at the top. You may use a full-size tactical scope with YHM 5-inch rail extension (YHM-9474) as shown to clear the rear sight protective wings. I call this arrangement "rail stacking," where two rails can be stacked to perform a special task. This setup will require a cheek riser for the buttstock.

Another excellent use of rail-on-rail combination to clear a fixed backup rear sight using a full-size scope is combining the YHM rail extension with Brownells flat-top riser. This setup is used with the Evolution tactical 10/22 equipped with AGP side-folding stock. Note the YHM raised, bull-barrel gas block. This same-plane gas block allows the use of the shorter same-plane, flip-up front sight (YHM-9627) as shown.

Another way of rail stacking to clear the height of a fixed-type backup rear sight with a full-size scope is to couple Brownells' AR-15/M16 flattop riser with RRA's high-rise cantilever scope mount, as shown on this AR-22 chassis.

Brownells' AR-15 flattop riser elevates the Bushnell Trophy to just the right height to clear the DPMS stand-alone, nonfolding rear sight. This riser is one of the best buys out there for this type of accessory. Note the custom lever safety I designed for this Supreme Match Rifle (SMR22), complemented by the Ram-Line 50-shot quad magazine system for an awesome 200 rounds of firepower.

CHAPTER 8

Custom Barrels and Muzzle Devices

Custom barrels and muzzle devices for the 10/22 have been around for at least three decades now, possibly since the 1980s. This is not surprising, owing to the fact that the 10/22 is not only the most popular model in its class but, undoubtedly, the most reliable magazine-fed semiautomatic rimfire rifle ever made.

The factory barrels are pretty accurate for the purposes for which they were designed: informal target shooting, plinking, small-game hunting, and action shooting events. For more serious applications, however, such as long-range hunting and rimfire rifle competitions, a match-grade barrel is definitely desirable.

Over the years, dozens of independent 10/22 accessory producers—including Clark Custom, Green Mountain, Volquartsen, Kidd, Yankee Hill Machine, Tactical Innovations, and Tactical Solutions (to name just a few)—have come up with some of the most accurate and sophisticated match barrels for the 10/22. These barrels are available in regular steel or stainless, in standard plain or fluted profile, and in a finned style patterned after the famous Tommy gun barrel. Some of the more expensive exotic barrels are made of lightweight aluminum with steel liners and carbon fiber, steel-lined barrels with a threaded front end to take AR-type muzzle devices. Realizing the demand for a serious competition-type 10/22, Ruger eventually introduced its Target model, featuring the heavy hammer-forged, spiral-design match barrel with recessed target crown. Complementing the match barrel is a factory-tuned target trigger that provides crisp, no-slack trigger pull. All in all, the Target model is a "match-ready" competition gun.

I selected the barreled action of the Target model for the RB Evolution Supreme Match Rifle (SMR22). This factory, match-grade rifle is very heavy, as it comes out of the box with a 20-inch bull barrel and laminated wood stock. Designed primarily as a benchrest rifle, the Target is definitely not geared for iron sights but strictly formatted for telescopic sights. With good quality optics and match ammo, this rifle is more than capable of 1-inch groups at 50 yards when fired on a bipod or benchrest.

In addition to the U.S.-made match barrels, a noted privately owned AR rifle manufacturer in Canada, Dlask Arms, also produces high-quality steel blued barrels for the 10/22 in plain or fluted, in heavy match-grade format. These are available in 8, 12, 16, and 18 inches in length, including a 12-inch finned barrel specially made for the Fabsports Thompson .22 kit. I am quite impressed with Dlask barrels for their excellent accuracy. The 16-inch fluted heavy barrel surprisingly outscored the Ruger 20-inch, hammer-forged, heavy match barrel in accuracy using a variety of economically priced, high-velocity .22 LR ammo at 25 yards. However, the Ruger 20-inch barrel delivers tighter groups at 50 yards and beyond. Dlask utilized the same top-quality barrel material it used for its AR-15 barrels. Despite their superb quality and workmanship, Dlask's products actually cost less than comparable match-grade barrels made in the United States.

These AR-15 compensators work extremely well in both the centerfire .223 and .22 calibers. When used with rimfire guns, these devices trap and accumulate heavy particles inside and are very difficult to clean. They are best left for full-bore tactical firearms.

Variations of muzzle devices available for AR-15/M16-type rifles that can be employed with rimfire tactical .22s. Flash hiders with elongated slots and open front ends work best with rimfire guns. They are easier to clean, with less lead buildup during prolonged shooting sessions.

Muzzle devices—such as a flash hider, compensator, brake, or suppressor (live or fake)—have become integral accoutrements for tactical firearms. On .22-caliber rimfire rifles, muzzle devices are more or less decorative, especially if the gun is already equipped with a heavy barrel. For diehard military-arms enthusiasts, however, a tactical firearm without a muzzle device is incomplete.

The market is currently flooded with AR-15 muzzle devices. They come in all shapes, sizes, and styles for many popular calibers ranging from .22 rimfire up to .50-caliber BMG and tactical shotguns. Custom 10/22 barrels already threaded to accept AR-15 muzzle devices are available from Yankee Hill Machine, Dlask Arms, and Tactical Innovations. Tactical Innovations has a wide variety of these custom-threaded barrels ranging from 5 1/2 inches to 18 inches in length. The shorter barrel (less than 6 inches long) is fluted and a standard part on its Cohort 10/22 pistol, whereas the 8-inch barrel is finned. This custom barrel is a perfect replacement for the Ruger Charger pistol. This barrel comes threaded and is ready to be attached. Note that rifle barrels under 16 inches are subject to federal rules and regulations in the United States, as well as possible state or local ordinances. According to National Firearms Act (NFA) restric-

tions, it is illegal to install (or even have in your possession) a short barrel on a model that originally came with a standard 16- or 18-inch barrel. Always consult with the appropriate authorities before attempting to install a short barrel on your 10/22.

I have done extensive testing of various AR-15 flash hiders and muzzle brake devices installed on the 10/22, firing various brands of ammo made by Winchester, Remington, Federal, and CCI. The results are quite interesting. Muzzle devices incorporating intricate cutouts, such as the Primary Weapons FSC556 and such multichamber designs as the DPMS Meculek compensator and YHM Phantom muzzle brake/compensator, work great with full-bore battle rifles but are not ideal for .22 rimfire cartridges. When used with .22s, these muzzle devices tend to filter lead particles coming out of the muzzle, and after few hundred rounds, the particles clog the interior of the device. The chambers and side ports become coated with lead, which is difficult to remove without using a special scraper that you have to build yourself from a soft metal, such as bronze or aluminum. Do *not* use a steel scraper, as it will damage the surface of the device.

The best AR-15 muzzle devices for use on the 10/22 are the ones with straight, elongated slots (bird

Close-up of the Volquartsen 10/22 stabilization module slip-on muzzle device designed primarily for bull-barrel models. The module features two rows of vent holes drilled around the body to stabilize bullet travel and muzzle control. The baked-on finish peels off quite easily on this device. Note the QD Versa-Pod tactical bipod installed on my custom mount.

Close-up muzzle view of the Spike's Tactical ST-SD-1 fake suppressor. This device is hollowed inside and shrouds the front end of the barrel to make it look shorter. It is ideally suited for full-size rifles and carbines with 16-inch barrels.

Fake suppressors are gaining in popularity among tactical gun aficionados who cannot own the real thing. These fine examples are made by (left to right) Tactical Innovations, Fabsports, and Spike's Tactical.

cage–style) and the spiral-slot design because they don't clog up even after many hundreds of rounds. These are mostly employed as flash suppressors rather than compensators or muzzle brakes. Just a few of these devices are the Carlson Mini Comp, Vltor VC-1, Smith Enterprise Vortex, DPMS A2 flash hider, YHM Phantom 5C1 and 5C2, Brownells Tactical flash hider, and Troy Industries Medieval flash suppressor. The open front-end design of these devices allows easy internal access, making them much easier to clean without the need for a special scraping tool. They also cost less than most high-end brands with special cutouts and multichamber designs.

For nonthreaded barrels, slip-on flash hider/muzzle brakes are available. Some work, and some don't. If you care more for appearance than actual effectiveness, there's nothing wrong installing a slip-on device. For bull-barrel types, Volquartsen has a slip-on muzzle device that actually works, the Forward Blow Stabilization Module. This device aligns the exiting gas so it surrounds and stabilizes the bullet for increased accuracy. It is made from lightweight aluminum and is available in silver and black finishes. Though this accessory is well made, the paint finish peels off quite easily as a result of the gas blast escaping through the portholes and bullet exit hole. A mil-spec hard-anodized finish would have been a much better choice than paint for a device that is exposed to a high heat stress. The module must be disassembled every 1,000 rounds and cleaned from the inside to remove lead-particle buildup.

Another type of muzzle device gaining in popular-

The Spike Tactical fake suppressor removed and the Meculek compensator substituted, showing the length of the Green Mountain 18-inch, match heavy barrel that is concealed by the fake device when installed.

ity among tactical firearms owners is the fake suppressor. These devices come in various styles and sizes.

A look-alike AK dress-up kit for a 10/22, such as the popular Krinker-Plinker carbine, features a full shroud that looks like a suppressor to conceal the portion of the barrel that extends past the forearm to give it a "shorty" look. The same arrangement was also employed on the GSG5 carbine to emulate the external format of the famed Heckler & Koch MP5SD silenced 9mm submachine gun.

Tactical Innovations currently markets a fake suppressor for .22-caliber guns, the FS22, which externally looks identical to the company's real NFA-registered suppressor. It is CNC-machined from solid aircraft-grade aluminum; coated in a durable, hard,

black-anodized finish; and laser-engraved for an authentic look.

Spike's Tactical offers fake cans for AR-type rifles in both 9mm and .223 calibers. The company's fake suppressors are available in two formats: the ET-1 and SD-1. The ET-1 is made for short-barreled rifles and adds 5 inches to the barrel's overall length, whereas the SD-1 shrouds back over the barrel, adding no length to the barrel. This accessory transforms a standard M4 carbine with 16-inch barrel into a "shorty" commando-style firearm, concealing the actual overall length of the barrel with the fake can. Spike's fake suppressors are machined from solid 6061 bar-stock aluminum and sport an anodized flat-black finish.

Although these decorative devices are well

made, they definitely need aesthetic improvements. They are very plain, and some of them are definitely long and heavy, depending on the length and diameter of the device. Ideally, in cases where an oversized fake can is desired, the body of the device must be turned by about half an inch in diameter, leaving only about half an inch of each end of the part untouched, thus creating a skeleton body. Thin aluminum tubing is then sleeved over the lightened section of the body to bring it back to its external dimension. The exterior tubing is then spot-welded in place so it cannot be removed to prevent unwanted alteration. This format will add authenticity to the overall profile of the device.

The hollowed section of the main body would make the device extremely light in comparison to the solid one-piece body construction being offered today. The shroud-over fake can, such as the SD-1 model, is very light since it is bored out to sleeve over the barrel to accommodate a barrel of up to 1 inch in diameter. To make the device even more aesthetically pleasing, the exterior tubing can be drilled with an alternate series of small holes to look like a ventilated barrel shroud or an oversized muzzle brake/compensator similar to the one employed on the famous German FG-42 assault rifle of World War II.

NOTE ON MATCH BARRELS WITH BENTZ CHAMBERS

Custom-grade match barrels are specialty items geared to produce maximum accuracy. There are many match-grade barrels available for the Ruger 10/22, and they are more expensive than standard barrels. Barrels that have been specified to have a Bentz-type chamber have a tighter dimension and shallower chamber to strictly accommodate match-grade ammunition. This type of chamber is more finicky with certain types of ammunition, which can cause extraction problems, especially after the chamber becomes dirty or clogged with powder residue after prolonged use.

If you are experiencing extraction problem with your match barrel, check the tightness of the chamber by following these simple steps.

1. Draw the bolt all the way to the rear and hold it open through the hold-open device.
2. Remove the magazine and make sure the chamber is empty.
3. Insert the cartridge you are using in the chamber with the muzzle pointed down. The cartridge should fall freely inside the chamber with minimal hesitation. If the cartridge is too tight to slide into the chamber freely, you have a Bentz chamber and you may have to select another brand of ammo to achieve a reliable operation.
4. With a cartridge in the chamber, invert the gun with the muzzle pointing upward. The cartridge should fall freely or at least be easy to pull out manually with minimal tightness.

Often, gun owners who are not aware of this problem change the extractor (even if the gun is still new), only to find out that the problem still persists. If the gun is intended solely for competition and will use only match ammo, there is no need to modify the chamber. However, if it is to be used as an all-around gun, it is best to buy a chamber reamer and convert it to factory specs. Once this modification is done, you can shoot virtually any type of ammo and expect reliability, even after many hundred of rounds have been fired without cleaning.

Brownells sells different types of chamber reamers for .22-caliber barrels. Find one that suits your specific needs and price range. I used the Manson Precision rimfire cartridge .22 LR match finisher reamer to fix the tight Bentz chamber on my heavy barrels after experiencing loading and extracting problems using a variety of .22 LR ammo. The malfunctions have ceased, and the gun works flawlessly. You need to change extractors only if the gun is old and has fired at least 10,000 rounds. I have an old 10/22 that has fired more than 25,000 rounds, and it still works like new. The only work done on it was replacing the extractor spring with an extra-power one and adjusting the hook of the extractor to compensate for wear. This extractor worked for another 7,000 rounds before it was eventually replaced. The 10/22 is truly an amazing gun.

CHAPTER 9

Bipod Selections

Bipod designs have come a long way when compared to their vintage counterparts. This very useful firearm accessory was originally designed for light machine guns and sniping rifles. In World War II, the use of the bipod was expanded even to long-barreled submachine guns being employed in the role of secondary light machine guns. In postwar-era conflicts, clip-on bipods became the norm on U.S. service rifles, while European models incorporated their bipods as a built-in accessory that neatly tucks underneath the

The latest generation of bipods made by CAA, Harris, and Versa-Pod compared to the vintage Vietnam-era clip-on types (seen on the lower right). A bipod is a wonderful aid to accuracy, particularly for long-range shooting and competition.

forearm when not in use. Former communist-bloc military arms normally fastened their bipods to the gun barrel itself and folded under when not in use.

Today's combat bipods are truly high-tech and perfectly complement the weapons they were designed for. This new generation of tactical bipods is well made, radically engineered, and user friendly. Manufactured in various forms and materials, the bipods are priced accordingly, from cheap all-plastic models under $20 to top-of-the-line machined aluminum/polymer/rubber combinations that fetch upwards of $250.

A very interesting variation of these new types of bipods is the Grip Pod. This innovative device is a unique combination of forward pistol grip and built-in pods. The legs spring out with the push of a button and retract neatly inside the grip when not in use. The advantage of this system is that it does not have the usual bulk of a bipod, thus leaving a lot of room on the weapon's mounting rail while providing the additional leverage of a forward pistol grip for maximum firing control in rapid-fire mode.

Dozens of superb combat bipods are available from some of the biggest names in the tactical accessory manufacturing sector. This lineup includes the Mod-Pod by Vltor, the AR-15/M16 Advanced Combat Bipod by Tango Down, the Versa-Pod Battle Pack by Keng's Manufacturing, and the 1A2 and Series S bipods by Harris. Grip Pod Systems introduced the original Grip Pod more than a decade ago. A more sophisticated variation was developed and manufactured in Israel and is currently marketed by Command Arms Accessories (CAA). You can see photos and descriptions of these marvelous bipods at Brownells website (www.Brownells.com).

Bipods used in the 10/22 tactical models presented in the book include the Harris Series S model. Harris is one of America's oldest innovators solely dedicated to bipods, and its brand is synonymous with quality in the bipod manufacturing industry. Its new Series S line is an improvement over the classic spring-loaded telescoping leg design so popular for many decades among hunters, snipers, and competition shooters worldwide. Bipods in the S series rotate 45 degrees for instant leveling on uneven ground. Despite the entry of more sophisticated and expensive combat bipod systems into the market, Harris bipods remain the favorite among hunters, competition shooters, and U.S. military snipers who demand maximum stability on a rugged platform at a reasonable price. The Harris is also available in the original 1A2 configuration, featuring a solid base and non-swiveling action.

There are several variations of Harris bipod adapters currently available, including those made by Harris itself. Two of the brands, from Yankee Hill Machine and Double Star, match the Harris bipod perfectly and are highly recommended for those who like the tactical format. These brands are very high-quality CNC-machined aluminum, have a nicely anodized finish, and are reasonably priced.

I also tested and evaluated the Keng's all-metal Versa-Pod for its overall performance, and the results were less than favorable. It is available in 12 models, including a "battle pack" version. The top-of-the-line model combines the highly adaptable Model 52 with three interchangeable sets of legs that let the operator configure the bipod for a variety of terrain and shooting applications. The bipod features a pan-and-tilt mechanism that allows it to pivot right, left, up, or down to compensate for uneven ground. While the overall system is sound and workable, it has one major flaw. Unlike the Harris Series S design that features side flat springs to support side-to-side movement for better operator control, the Keng's design leaves its movements unsupported. Thus, the weight of the gun causes the bipod to flip freely from

The Harris bipod installed and in deployed position, ready for long-distance shooting. Note the Ram-Line quad 30-shot magazines (no longer made) that provide an awesome 120 rounds of firepower.

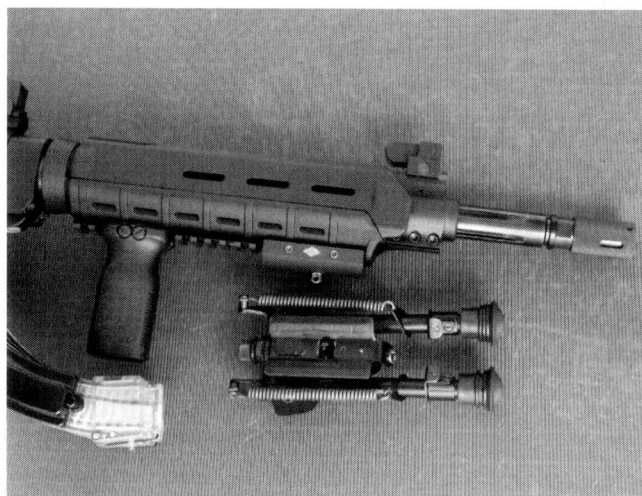

YHM makes an excellent bipod adapter for the Harris bipod. The YHM adapter is seen installed on the Magpul mid-length forearm bottom rail. Note the RVG vertical grip, also made by Magpul, complementing this high-performance AR-22 custom tactical rifle setup.

side to side, making it sloppy and less controllable for the operator, especially if the gun is heavy. These bipods can also be mounted on rifles equipped with Versa-Pod adapters, which are sold separately.

Among the current breed of combat bipods, I have found the new CAA tactical bipods to be the best choice in terms of advanced design, quality, and price. These bipods feature lightweight overall construction, a quick-release push-button leg extension, and stainless steel legs that contain a spring-loaded mechanism. These newly redesigned legs provide increased stability. CAA's new generation of bipods is offered in two versions: a Picatinny rail system and a

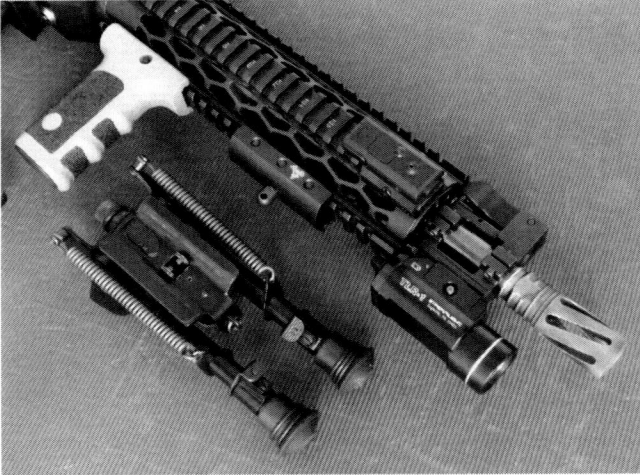

Another company that makes an excellent bipod adapter for the Harris bipod is Double Star. Its adapter is shown here installed on the YHM Specter Diamond free-float forearm. Though the two are similar in design, the Double Star adaptor uses a stamped clamp and three screws, whereas YHM's clamp is all aluminum with double retaining screws.

side-mounted rail system. The Picatinny design mounts onto the tactical rail in seconds. The bottom of the bipod features an additional rail, which is quite helpful with a personal defense weapon (PDW) that uses a mini-forearm rail to permit the mounting of a forward grip, laser, or light. The new recessed design allows the legs to be mounted higher, giving a lower, more stable center of gravity. The polymer-covered pods provide a nonslip grip when pulling the legs from the folded to deployed position.

The CAA BSRS is my top choice among the latest tactical bipods. This unique bipod has all the advantages offered by CAA's standard Picatinny rail bipod but also features individual mounting bases that easily attach and detach in seconds from the side rail of a forearm. The advantage of the side-mounted system is it lowers the bipod's center of gravity, resulting in a very stable platform compared to bottom-mounted types. This arrangement leaves virtually the entire length of the forearm's bottom rail free for whatever tactical accessories the operator wishes to install. The side mounts can also be left installed in place without the legs if the operator wishes to lighten the setup for shoulder firing. The legs can be installed in seconds if the weapon is to be fired in the prone position.

This radical AR-22 sniper simulator features the CAA/EMA Tactical Picatinny bottom-mounted detachable bipod. This rugged bipod has fully enclosed mechanical components, which means they are less affected by adverse climatic conditions. A unique feature of this bipod is the quick-detach legs option. The legs also can be folded to the rear or to the front when not in use.

This impressive-looking FS556 SBR features the CAA side-mounted bipod installed on the side rail of the YHM carbine-length, customizable free-float forearm. Note the fake suppressor used as an alternate muzzle device to the YHM Phantom compensator/brake.

Tango Down developed and produced a variation of the side-mounted bipod, called the Advanced Combat Bipod, which was noted earlier. The Tango Down version has the base of the pods split into halves and formed as an integral part of each leg for a much simpler arrangement. However, the Tango Down unit costs about $100 more than the CAA model, making it one of the most expensive combat bipods on the market.

CHAPTER 10

Best High-Capacity Magazines and Loading Apparatuses

A high-capacity magazine is definitely the one accessory that puts the finishing touch on a 10/22 tactical package. Unfortunately, despite the expiration in 2004 of the decade-long Federal Assault Weapons Ban that prohibited the possession of high-capacity magazines for both short and long guns, citizens in a handful of states and cities are still living under the umbrella of this controversial "crime" bill, where magazine capacity is still limited to a specific number of rounds. Check all state and local laws about magazine restrictions before you acquired a high-capacity magazine.

The first of the high-capacity magazines ever offered for the 10/22 back in the early 1980s was the Condor 25-shot banana-type magazine made by Eaton. This magazine worked quite well, but it had one major flaw: the small part at the back of the housing that engages the magazine catch broke too easily because of the fragile plastic material from which it was made. Butler Creek remanufactured the Eaton magazine in the 1990s in a much refined form with an unbreakable plastic housing and magazine lip called Hot Lips. Butler Creek also offers a model

Banana-type, high-capacity plastic magazines for the Ruger 10/22 have been around for more than three decades, starting with the 25-shot Condor. This original magazine was soon followed by many more variants, including those made by Ram-Line, Eagle International, Butler Creek, and Tactical Innovations. Ram-Line's compact double-stack magazines were revolutionary improvements over the single-stack design.

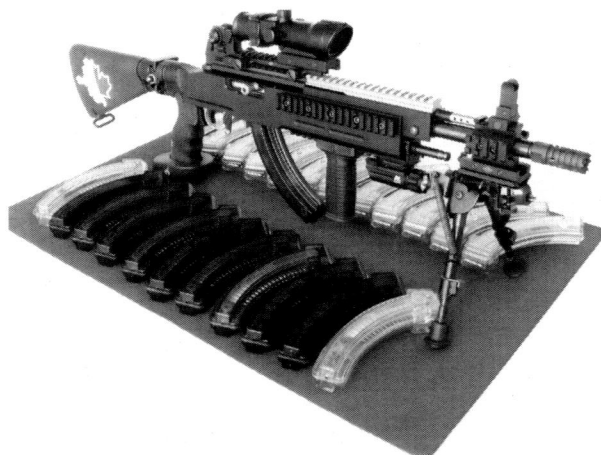

Butler Creek produces arguably the most reliable 25-shot polymer magazine with its Steel Lips version. The more economically priced plastic lips version, Hot Lips, is also reliable, but the edges of the lips are paper thin and easily deformed over time. These magazines are available in clear and smoke colors.

with rugged steel lips, and I consider this version, called Steel Lips, the best of the new generation of banana-type magazines available today.

Another remarkable polymer banana-type magazine was Eagle International's 30-shot version, released back in the mid-1980s. I have two of these magazines made by the original company based in Arvada, Colorado, and they still work after 30 years of regular use. A new Eagle International company located in Hazen, North Dakota, revived the manufacture of all the vintage 10/22 and Mini-14 accessories produced by the Colorado-based company, including the 30-shot banana magazine. Now available in black, smoke, clear, and blaze orange, the new Eagle magazine is well made and reliable, and costs only $10. I consider this magazine the best value around.

Ram-Line was among the early pioneers in the design and manufacture of high-capacity magazines for rifles and handguns. Its revolutionary polymer folding stocks for the Mini-14 and 10/22 put Ram-Line in the forefront of magazine design. Combining its patented indestructible aerospace polymer materials with its innovative and compact double-column magazine system for the 10/22 set Ram-Line apart from its competitors. This double-stack magazine was truly a breakthrough in rimfire-magazine design. They were produced in 15-, 20-, 30-, 40-, and 50-shot capacities. Early single-stack, polymer, banana magazines produced by Ram-Line had a 30-shot capacity. Ram-Line also manufactured a limited-production 25-shot magazine, the Truncator, which was developed strictly for use with cartridges utilizing truncated bullets, such as the Remington Yellow Jacket. This hypervelocity ammo normally causes high-capacity magazines to jam, primarily because of its bullet configuration.

I still have a dozen of the original Ram-Line magazines from the 1980s, both single- and double-stack. Some of the double-stack versions still work, and some don't. The double-followers tend to jam each other when the lubrication dries out. Silicon oil usually solves the problem, but not all the time. Spring fatigue is another problem in these 30-year-old double-stacked magazines. The 30-shot single-stack banana-type magazines, however, still work as well as when they were first manufactured. I consider

these Ram-Line 30-shot magazines the very best in their class—they can nearly duplicate the overall reliability of the factory rotary-type magazines and yet also yield three times the firepower.

Shooters Ridge is now producing a spin-off of the 30-shot Ram-Line double-column magazine. Shooters who have tried these Shooters Ridge magazines have posted mixed reviews, with the majority being negative. Such complaints prompted the manufacturer to include a pamphlet with each new magazine sold on how to correct malfunctions, including minor fitting to the gun itself. Obviously, there are some dimensional problems with the design, since most magazines do not require this much adjustment on both the magazine and gun to make them work. Unfortunately, I was unable to acquire and test a magazine from Shooters Ridge (including its newly released single-stack version), so I cannot comment on their reliability or quality.

In addition, Shooters Ridge also designed and produced a special loader for its magazines called the 10/22 Magazine Loader. The universal loader features a crank-operated thumb piece to activate the cartridge pusher. Ram-Line pioneered magazine-loading systems for the 10/22 in the 1980s with its autoloader device following the introduction of its double-stack

In addition to the development and manufacture of sophisticated high-capacity banana-type magazine systems for he 10/22, Ram-Line also introduced a matching "autoloader" device, shown here with its compact 20- and 30-shot double-stack magazines. This device also works with single-stack, banana-type magazines.

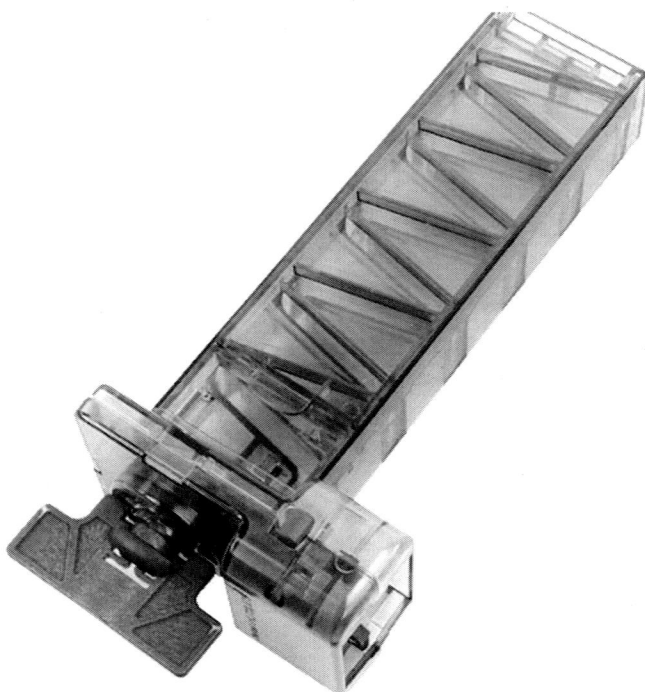

The Butler Creek magazine loader is among the best of the banana-type high-capacity magazine loaders available for the 10/22 today. It will work in both single- and double-stack magazines. The loader is seen here with Butler Creek 25-shot and Ram-Line compact 20-shot magazines.

Shooters Ridge recently reintroduced the original Ram-Line double-stack magazine system in the 10/22 tactical gun market. To complement these magazines, Shooters Ridge also markets a loading device that features a unique zigzag conveyor system. The hand-crank operating system is similar to the Butler Creek design (right), but the conveyor system is original.

magazines. The Shooters Ridge loading device incorporates a cartridge conveyor system that works on a zigzag principle, while the Ram-Line system employed a series of vertical conveyors that taper down to a single line at the bottom of the loader. The Ram-Line loader utilized a push-button mechanism to activate the loading system. Unfortunately, none of these high-capacity magazine-loading devices work 100 percent of the time. This can be attributed to the varying rim diameters and overall length dimensions of the diminutive cartridge itself.

Other derivatives of the classic banana-type magazines sprang up as soon as the ban on high-capacity magazines was lifted. Among the new releases, one entry is worth mentioning, primarily because of its unique features. Tactical Innovations' TI25 magazine series is externally similar to the Butler Creek 25-shot magazine. The TI25 models are available with machined-aluminum or polymer housing. A unique

feature of these magazines is their two-piece construction, which allows them to be separated in order to access the follower and spring for a thorough cleaning after prolonged use. There are also four screws at the top of the magazine ledge that can be adjusted to minimize the fore and aft play when the magazine is seated to the gun, which, according to the manufacturer, improves feeding.

Other manufacturers explored alternate ways to design and manufacture ultrahigh-capacity magazines for the 10/22. For example, the early Sanford 50-shot is an awesome-looking drum magazine that predates other 50-shot magazines for the 10/22. Despite its impressive design characteristics, the Sanford drum did not live up to its billing. It was a total failure in terms of reliability. The quality of materials used is very poor, the clear plastic casing breaks easily, and, instead of steel, the magazine lip is made of cheap, soft-cast zinc oxide (alloy) material normally used on gas pistols and replicas. The winding mechanism and clock-type spring appear to be sound, but the magazine follower is a long way from perfect, and the magazine lips are out of spec. The lips release the cartridge too soon, causing it to jam between the bolt face and chamber mouth of the barrel. The winding mechanism also tends to hang up erratically, thereby

The Sanford 50-round drum magazine was the first to enter the ultrahigh capacity magazine market for the 10/22 in the early 1980s (left), followed by the Pro-Mag version (middle), and more recently the Black Dog (right), the most reliable among the three.

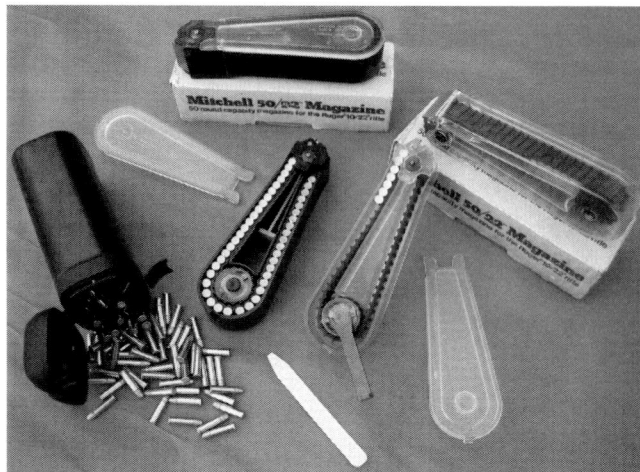

Shortly after the entry of the Sanford 50 in the high-capacity magazine market, Mitchell Arms introduced its 50/22 Teardrop magazine system, which was available for both the 10/22 and AR-7 Explorer carbine. The 50/22 featured a unique belt-loading system and two-tone plastic casing. It was not a commercial success because of its unreliable operation.

leaving no round in the magazine lip, resulting in an empty chamber. Although the now defunct AMT company (which produced stainless steel versions of the 10/22 and Mark I pistols) adapted the 50-shot Sanford for its AMT Lightning rifle to compete with look-alike imports, it was not a commercial success.

In the early part of the new millennium, Pro-Mag produced a modernized version of the Sanford drum. The Pro-Mag drum uses a separate ratchet-type winding mechanism, whereas the original Sanford design featured a back window through which the winding mechanism could be operated by finger. The overall design of the Pro-Mag version is excellent, and the plastic casing appears to be unbreakable.

There are, however, several flaws with the Pro-Mag design. First, Pro-Mag made the same mistake as on the Sanford magazine: using soft metal instead of steel in the magazine lip area. Another flaw is the very short protrusion of the round metal in front of the magazine that connects to the receiver. When the gun is fired, the magazine tends to disengage and fall off the gun, since this part of the magazine is barely engaging the gun. The magazine connection to the receiver is loose, and it tends to rock back and forth quite easily when it is seated to the gun. The feed

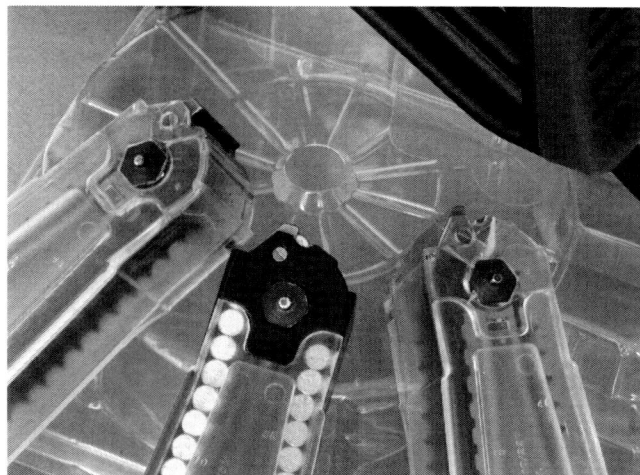

Close-up view of the vintage Mitchell Arms 50/22 Teardrop magazine (center), seen between the latest MWG versions with cracked upper casings. MWG used fragile plastic material in its magazine body, while the Mitchell remains intact after more than 30 years.

ramp of the magazine is out of spec, and the breech face of the bolt that pushes the cartridge toward the chamber hits the magazine feed ramp and deforms it. The magazine produces acceptable reliability if the nonshooting hand supports it while the gun is being fired to prevent it from falling off. When not supported by the hand, the downward racking action of the magazine as the bolt moves forward tends to jam the cartridge nose against the rim of the chamber mouth as it comes out of the magazine. As you can see, the Pro-Mag 50 is far from perfect. It still needs further refinement in the above-cited areas of concern.

Another recent entry into this tactical arena is the Black Dog 50-shot polymer magazine for the 10/22. This magazine looks promising and is very well made in comparison to the Sanford and Pro-Mag versions. The Black Dog drum appears to have borrowed some features from the Sanford, such as employing the back windows to access the winding plate and its clock-type spring mechanism. Its inherent reliability can be attributed to the features borrowed from the Italian-made Bingham PPS-50 incorporating plastic dummy cartridges as part of the magazine follower system. The PPS-50 is possibly the finest 50-shot drum magazine ever made, and Black Dog certainly made an excellent choice in employing its design characteristics. In addition to the 10/22, Black Dog also offers 50-shot drums for the AR-15 with Atchisson-type conversion kits, including the new version of the Italian-made Bingham PPS-50 called PUMA.

The most unique 50-shot magazine system from the 1980s was the Mitchell Teardrop magazine. The Mitchell magazine features a belt-loading mechanism that rides on a sprocket apparatus mounted on the lower part of the magazine casing. The magazine lip is made of cast steel, as is the factory part. The housing consists of two-piece, unbreakable plastic construction. The main body is black with a built-in belt hook, while the clear back (cover) allows the operator to easily see the number of loaded rounds in the magazine. The two specimens I bought 30 years ago are still intact today, except for the small, built-in tab at the bottom of the cover that locks the two-piece casing together. The tab broke eventually after repeated assembly and disassembly of the cover, which is required during the loading process. This is the weak area of the design.

As to reliability, the original Mitchell did not meet acceptable requirements. The main culprit is the magazine lip design. Although, Mitchell made the part in steel, it did not duplicate the exact pattern of the Ruger magazine lip, possibly for fear of patent infringement of the factory rotary-type magazine. This led to a variety of malfunctions, including cartridge nosedives, jams, and ejection problems. When I substituted the original factory magazine lips for the Mitchell part, the vintage Teardrop magazine worked flawlessly with most brands of ammo. Just like the Sanford drum, the Mitchell magazine failed commercially, primarily because it was not reliable. In both cases, the manufacturers failed to recognize the main culprit: the feed-lip design. Fixing this simple flaw would have made this magazine one of the most reliable and unique 50-shot magazines ever made for the 10/22.

Since the Federal Assault Weapons Ban sunsetted in 2004, many of the prohibited accessories and high-capacity magazines for assault-type weapons are once again being produced. The MWG Company took over production of Mitchell 50-shot magazines. The MWG version works extremely well, simply because MGW made the magazine lip exactly as Ruger had made it. However, as reliable as the new MWG version is, it has a fatal flaw: the plastic material used in its two-piece magazine housing is very fragile, just like the material used in the original Sanford 50-shot drum. The two specimens I bought not too long ago already have a crack around the magazine cap nut that controls the tension on the upper half of the belt. The bottom of the magazine housing itself suffered multiple cracks, rendering the magazine totally unserviceable, after just about 500 rounds loaded on each. At $80 apiece, this was a total waste of money. If MWG changes the housing to a better plastic material that won't break as easily, I will buy the magazines again and recommend them to others. For now, they work, but they don't last very long.

In April 2011, Ruger finally introduced a genuine fact-25. This new 25-shot banana-type magazine features the same correct angle-feed lips that deliver the unsurpassed reliability of its original 10-shot rotary magazine. The BX-25 features stainless steel feed lips and a stainless steel "constant force" spring. Its anti-tilt follower is injection-molded from Celcan—a high-velocity polyacetal. The magazine body halves are injection-molded from durable gas-filled nylon.

According to Ruger literature, the BX-25 has

proven to feed as reliably as the standard 10-shot rotary magazine with a wide variety of ammunition, but many shooters report experiencing malfunctions when using subsonic ammunition or ammunition with a heavy coating of bullet lube. The BX-25 requires no lubrication. It features two halves in its body construction, which are secured by two screws (top and bottom). This allows easy disassembly for cleaning the internals after prolonged use, similar to the TI25 banana-type magazines.

The Ruger BX-25 appears to be very well made and has received excellent reviews on its reliability from consumers who have tried it. Unfortunately, at the time of this writing, I was unable to get one for testing, so my comments and evaluation will be reserved for later.

The Ruger BX-25 is the company's long-awaited high-capacity magazine for its popular Ruger 10/22 carbine. It was an instant success when finally introduced.

CHAPTER 11

Building the RB Supreme Match Rife (SMR22)

The RB Evolution Generation One (or Gen-1) is one of the earliest of the modular stock-chassis systems for the 10/22 that can take an AR-15 stock (fixed or collapsible), pistol grips, and backup iron sights. When introduced, it was considered radical since there was nothing to which it could be compared. Years later, the Nordic AR-22 kit and Fabsports FS556 entered the market, relegating the Evo to the backseat. This led RB to redesign the Gen-1 chassis kit and come up with the upgraded Gen-2 model.

The Gen-2 is basically a stripped-down Gen-1 kit. The platform was shortened past the stock screw, thus eliminating the usual forearm altogether. In its place was substituted a removable one-piece, free-float forearm adapter that is connected to the front of the cut-off area by cross screws. Despite the new features of the Gen-2 design, it was no better than the Gen-1 in terms of aesthetics. In fact, it is possibly less attractive than the original Evo, which had a distinctive style of its own.

By today's standard, the Evolution modular stock system for the 10/22 can be rated as plain and unattractive. Its slab-sided design is a bit oversized when measured from the bottom of the foregrip to the top of the hand guard. Its overall size makes it heavy when compared to some of the latest entries in custom platform competition. However, this does not mean that the Evo cannot be improved—as long as the builder knows exactly how to do it. By applying a few cosmetic modifications (much like makeup on women), you can turn these plain-Janes into stunning tactical firearms. Since the action of the gun rarely needs additional refinement because of its proven reliability, the primary focus in this particular gun project is an aesthetic overhaul. Of the three chassis systems featured in this text, the Evo kit is the one that really needs the most work in order to compete with the updated, sophisticated 10/22 modular platforms in today's market.

Not everything is negative about the Evo. Except for its oversized, heavy overall construction and plain appearance, the RB platform is very well made and nicely finished. In fact, its velvety black-anodized finish is much better than the finishes on the Nordic and Fabsports kits. Although the chassis is designed to accommodate AR-15 furniture and pistol grip, the overall aesthetic of the Evo platform simply lacks a tactical touch for diehard military weapons enthusiasts. On the other hand, its massive solid platform will attract the more serious competitors who want a simple and easy-to-maintain stock system that can keep the action of the gun steady when being fired. The Evo's size and weight are advantageous here because they minimize muzzle lift and enhance accuracy and tighter grouping at longer ranges. For shooters who want to move beyond a simple setup, the Evo can be made into a world-class tactical marksman's rifle—the Supreme Match Rifle, or SMR22—when paired with the right accessories.

The following parts and accessories will be required to improve the overall aesthetics and handling characteristics of the Evolution SMR22.

1. Hahn Precision low-profile riser. This genuine AR-15 accessory is a one-piece rail cut to the military-standard 1913 Picatinny rail that matches the look and feel of a standard A4 flattop receiver. It adds half of the height when used on an AR flattop model for proper cheek weld with optics and tactical sights. In the Ruger .22 carbines, this part is used in conjunction with the factory-supplied scope mount to make the Evo stock a close facsimile of an AR flattop format, thus improving its military-style profile once the AR buttstock and pistol grips are installed. The cutoff (nonfunctional) AR rifle charging handle is assembled to the rear of this part and secured by a single screw from the underside. The Hahn riser will require drilling and tapping for a #6-32 pan-head machine screw at the rear.

2. Double Star AR-15 charging handle. This part complements the Hahn Precision accessory to maximize the Evo's AR-15 flattop profile complete with nonfunctional charging handle assembly. For looks, only the head of the charging handle (complete with the spring-loaded latch, also is required. The modified charging handle assembly will be connected to the rear underside of the Hahn Precision flattop riser by a 6-32 pan-head screw. The front-bottom end of this modified part will require an inclined radius to conform with the shape of the 10/22 receiver at the top rear, where it will be seated. This clearance cut will allow the Hahn riser to slide freely into the factory scope mount, complete with the AR-15 charging handle on top of the receiver. The head of the assembly screw for the charging handle must be seated flush to the countersunk and underside radius of the charging handle so it will not impede or scrape the surface of the receiver during assembly. (Refer to the workman's drawings on page 75 for dimensions required in the modification of the charging handle assembly and Hahn Precision riser.) (**Note:** A used charging handle assembly is ideal for this purpose since it will be cut off and modified. Gun shows or gunsmiths are the best source for used parts, and they normally are a lot cheaper than new ones. Among the most economically priced AR-15 charging handles are the Double Star and DPMS brands. At the time of this writing, they cost about $18 at Brownells.)

3. Evo scout rail. This is RB's optional 11 1/2-inch rail. This long rail is assembled at the top of the hand guard. The overhanging portion of the rail will normally ride over the receiver for mounting backup iron sights and optics. In the match rifle setup, this part was reverse-mounted with its extended front end riding over the gas block. This unique arrangement provides the competition gun an extra-long sight radius, which can deliver maximum accuracy with backup iron sights. The long space between sights yields a lot of room to accommodate various sizes of optics and other accessories.

4. Double Star .937 Picatinny gas block. This part, precision-machined from high-grade 6061 T6 aluminum, has an integral military-standard 1913 Picatinny rail on the top side. The gas block is secured to the barrel with a pair of hex-head screws on the underside. The rear-top section of this part will require a square notch to act as a rest for the front end of the scout rail.

5. DPMS .936 four-rail gas block. This part can be used as an alternate and functions similarly to the Double Star accessory for those who may prefer a four-rail option rather than a single-rail one. Some modification will be required to seat the front end of the RB Scout rail. (**Note:** Both the Double Star and DPMS .936 gas blocks will require wrapping a 4 1/2 x 1 1/2-inch aluminum pop can around the barrel to act as a sleeve to tighten the loose fitting.)

6. Falcon Industries 2-inch AR-15 hand guard rail. This accessory is mounted at the bottom of the Double Star gas block by two #8-32 hex screws, thereby adding an extra mounting rail. Installation of this part will require drilling and tapping the gas block. This is not required when using the DPMS gas block due to its four-rail feature.

7. YHM-9473 6-inch rail extension. This part is assembled to the Falcon mini-rail in reverse order. The front-end underside section of the YHM rail will require a 3/8-diameter half-round groove to seat the tip of the Versa-Pod mounting bayonet at the front end of the forearm. A Dremel tool sand-

ing band works perfectly for cutting the necessary clearance groove to the YHM accessory rail. (Refer to the workman's drawings on pages 76–77 for dimensions for the bottom rail assembly and versa-pod and bayonet mounting.)

The modification and installation of the above components are the most involved aspects of this particular *project. Once all these extra steps are completed, the next step for the gun's completion is basically straightforward and easy to follow.

Listed below are additional parts and accessories that will be required to complete the Evolution Ultimate Match Rifle (UMR) using the Ruger 10/22 Target model as a starting point. This top-of-the-line .22 features Ruger's exclusive hammer-forged, spiral-design heavy barrel with recessed target crown. The SMR22 described in this chapter is an enhanced verison of the UMR22.

8. Evolution stock—Gen-1.
9. RB scout rail.
10. RB 7-inch rail. Two are required for side-rail installation.
11. YHM-9464-7. A 7-inch rail for customizable forearm is to be installed at the bottom of the foregrip.
12. YHM-643K Solid A2 rear sight.
13. YHM-638 (bipod adapter).
14. ACE ARRB (AR-15 receiver block).
15. ACE AKFX-FS. This is an FAL-type skeleton stock complete with folding mechanism.
16. CAA FALC. This is an FAL rifle cheekpiece.
17. CAA OPSM. This is a one-point sling mount.
18. CAA BSR. This is an 8–12 inch side-mounted bipod (alternate).
19. CAA CSR3 30mm scope rings.
20. Magpul MOE AR-15 pistol grip (dark earth).
21. Tapco Fusion short vertical grip (dark earth).
22. Brownells AR-15 flattop riser.
23. Brownells CAR-15/M4 flip-up tactical front sight.
24. Volquartsen stabilization module (black).
25. Optics. Leapers UTG 30mm SWAT 3-12 x 44 full-size AO mil-dot RGB EZ-Tap scope.
26. Magazines. The Evo platform will accommodate all types of 10/22 magazines.

IMPORTANT NOTE ABOUT ASSEMBLY TOOLS

Although most manufacturers supply the necessary hardware for the installation of their product lines, some don't. Make sure you have all the tools to install or modify the accessories before you begin shopping for parts and accessories. Study each project carefully and select the model you can build with the tools you already have. An elaborate gun format will require some machine work or special tools. The services of a gunsmith or a machinist may be required for barrel turning and threading, as well as lathe operation (for more advanced projects) if you cannot do these tasks yourself or have no access to metal-cutting machines. In assembling the various projects guns in this book, I used the following tools:

1. Smith Enterprise armorer's wrench or DPMS multitool
2. DPMS M4 stock castle nut wrench
3. YHM forearm wrench to tighten forearm jam nut
4. Set of Allen wrenches
5. Gunsmith screwdriver set (optional)
6. Snap ring pliers
7. Dremel tool with flex-shaft attachment and set of grinding tools
8. Portable drill press
9. Set of drills, taps, and dies

PUTTING THE EVOLUTION SUPREME MATCH RIFLE TOGETHER

The following steps show how to assemble and disassemble all the necessary components to create the Evolution Supreme Match Rifle (SMR22) after all the necessary modifications of the aforementioned parts and accessories have been completed.

1. The Evo stock and matching hand guard are normally assembled parallel at the front. I found the size of the forearm a bit short and bulky, so I made it appear longer and sleeker by simply extending the front-end section of the hand guard one hole forward. To start the assembly, install both the side rails in their respective locations via the rear assem-

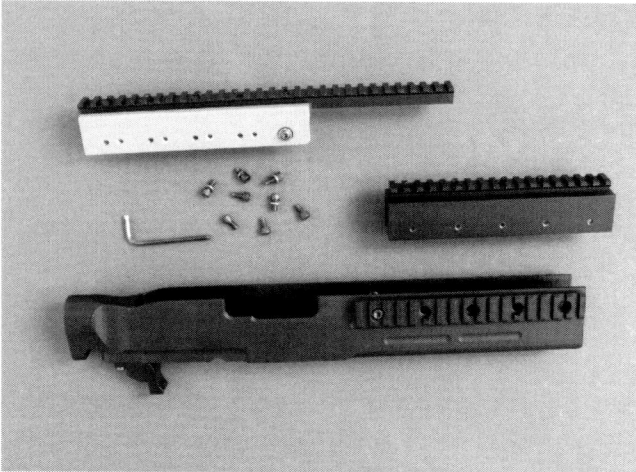

The first step is to install the scout rail to the top of the matching Evolution hand guard in reverse order, as shown on the top cover. You may also install the side rail to the Evolution stock using just the rear screw, as shown. The rear screw will function as a retainer for the side rail during routine disassembly.

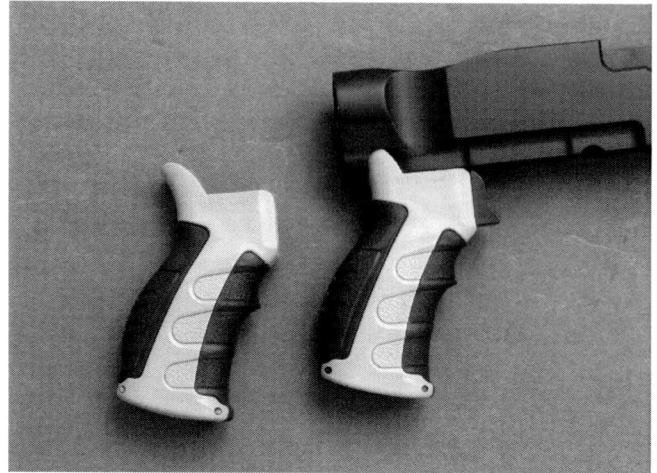

When using a beavertail-style grip, such as the CAA UPG16 or Magpul brands, the top portion of the beavertail will require fitting (as shown in this comparison photo) to fit the radius cut of the Evolution stock behind the grip seat. Trimming must be done gradually to achieve a seamless and perfect fit.

bly hole using a longer socket-head screw of the same thread. You will need a matching nut for this screw since you have moved forward the hand guard that has the matching threaded hole for this screw. When you assemble the side rails, make sure that any portion of this stationary screw that is protruding over the nut is cut flush so it will not interfere with a bull barrel when the action is assembled to the stock.

2. The next step is to assemble the YHM 7-inch customizable rail at the bottom of the forearm. You will notice that this rail is much thinner than a standard AR-15 bolt on this type of forearm rail. The reason I selected this particular rail instead of RB's own 7-inch rail was to avoid adding more bulk to its already oversized forearm (measured from top to bottom). Both the rear and front ends of the rail require threaded holes to accept #6-32 machine screws. The predrilled and tapped screw holes at the bottom of the forearm allow the screws to pass freely, so no additional modification is needed. However, the rear of the forearm near the stock screw requires one small hole to be

drilled to clear the #6-32 screw that connects to the rear of the YHM rail.

The front screw secures the assembly of both the Versa-Pod bipod mount and the front end of the YHM rail to the stock. The assembly screw of the YHM rail at the middle is assembled from the bottom of the rail via another predrilled hole of the forearm and then mated to a matching nut inside the stock. Both the front and rear screws are assembled from the inside of the stock and connected to the threaded holes of the YHM rail at the bottom of the forearm. Cut any protrusion of the assembly screws on the surface of the rail flush so that they will not interfere with the installation of accessories. (**Note:** It may be necessary to place a pad, fabricated from an ordinary poly-binder and measuring 7 inches long by 5/8 inch wide, between the rail and forearm during assembly to elevate the part slightly in order to accommodate all types of forward grip, which requires more space below the forearm during assembly. The plastic binder can be cut with scissors, while the three screw-location holes can be punched out with a .22 shell.)

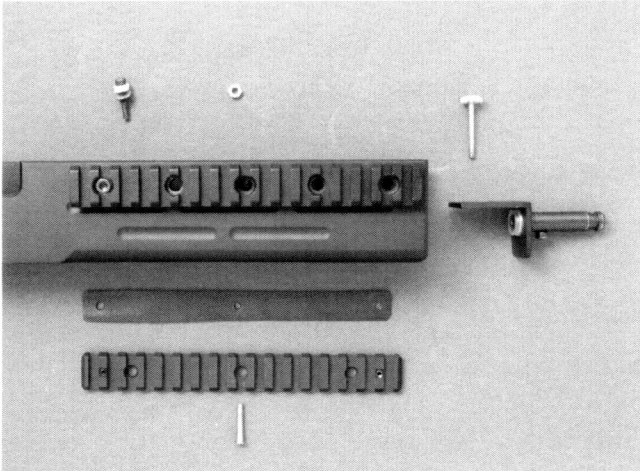

The completed custom Versa-Pod bayonet mount is seen in front of the forearm. The 7-inch YHM forearm rail and matching spacer (cut from plastic binder) is seen below, ready for installation in their proper order.

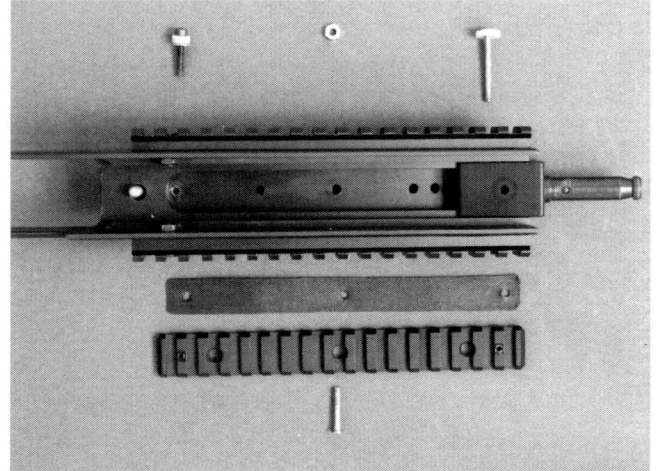

Top view of the Evo chassis showing the Versa-Pod bayonet mount placed on its seat ready for final assembly. The bottom rail and matching spacer is shown at the bottom. **Note:** The front and rear assembly screws will be assembled from the top while the center assembly screw will be assembled from the bottom and tightened by a matching hex nut.

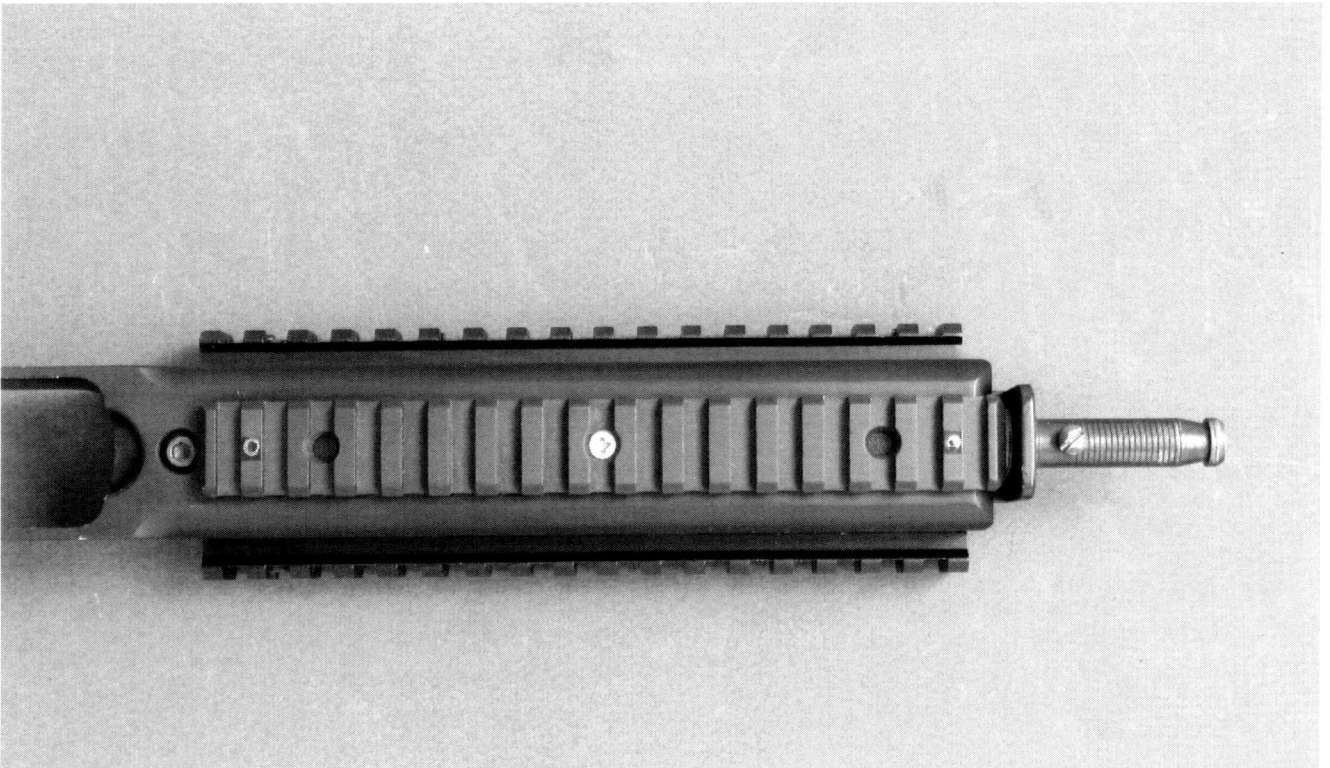

Bottom view showing the fully assembled rail and Versa-Pod bayonet mount.

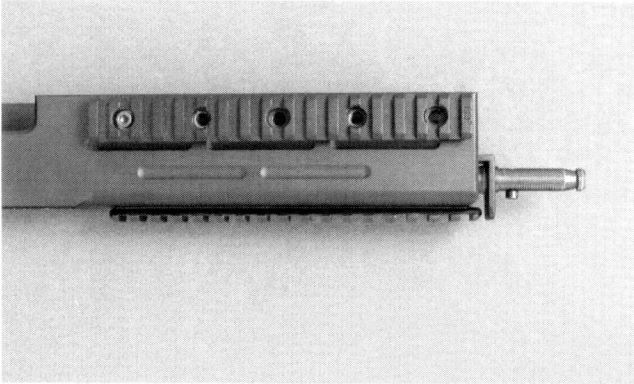

Side view showing the fully assembled bottom rail and Versa-Pod bayonet-style mount. Note the plastic spacer placed between the rail and underside of the Evo chassis, raising the rail slightly to accommodate various forward grip dimensions.

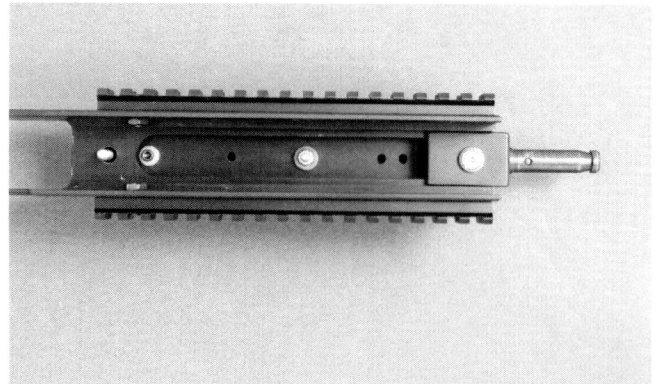

Top view showing the fully assembled Versa-Pod and bottom rail. This is a clean professional setup that RB Precision should offer as an option to their Evo stock system.

3. Assemble the scout rail to the matching hand guard.

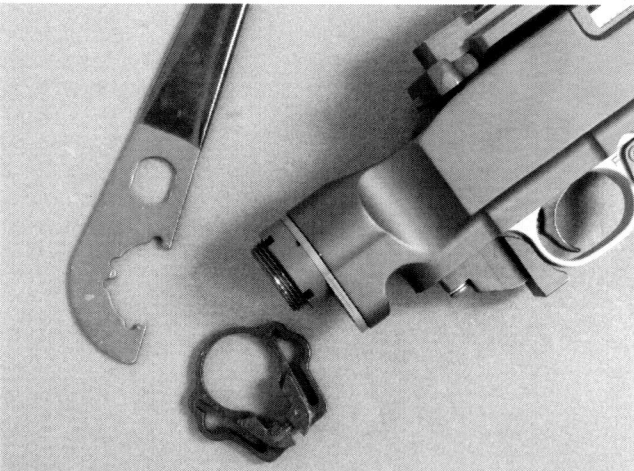

View showing the ACE AR-15 receiver block assembly installed at the rear of the Evo chassis. You will need an M4 stock castle nut wrench for this operation. Note the protruding threaded area of the receiver block to accommodate the CAA One Point Sling Mount (OPSM).

View showing the CAA OPSM installed. During this installation process, ensure that the rear end of the receiver block is flush or slightly protruding (.025–030 inch) above the OPSM to allow the unimpeded movement of the ACE folding-stock mechanism.

4. Assemble the ACE AR-15 receiver block assembly to the rear of the stock. (You will need a castle nut wrench for this installation.)

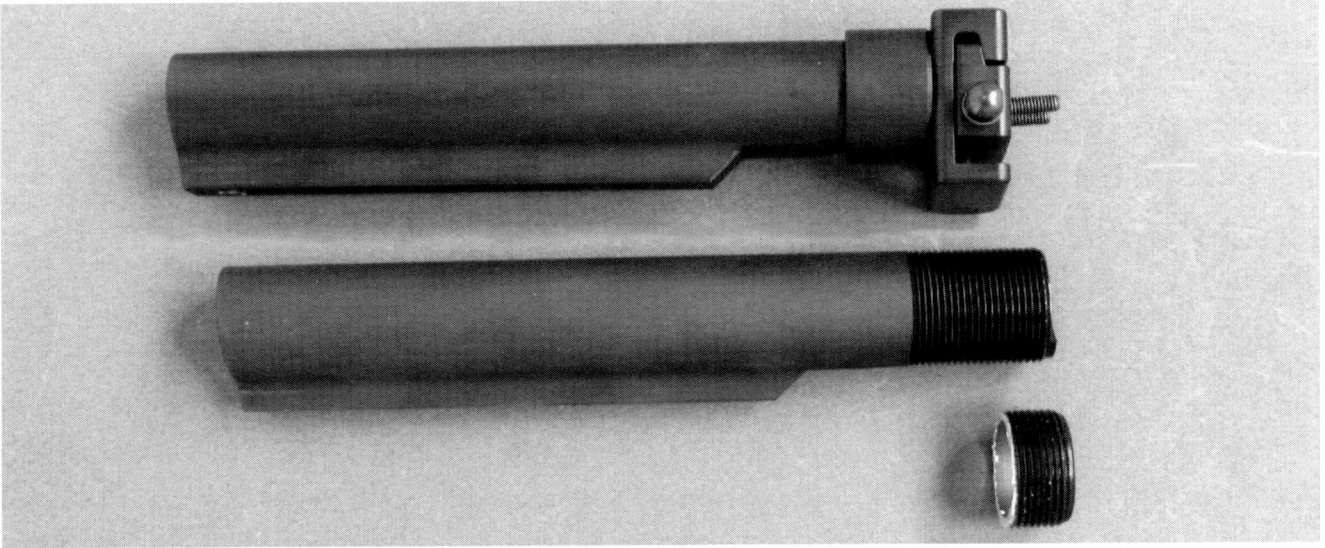

To enhance the aesthetics of the folding-stock assembly, reduce the oversize length of the M4 stock tubing as shown (bottom) and eliminate the unsightly threaded section of the tube showing during final assembly (top). Prior to assembly of the tube, use blue LocTite to the ACE part to prevent the stock from shifting once centered.

5. Install the CAA one-point sling mount over the castle nut. There must be at least .035 inch of the ACE receiver block protruding over the CAA mount to allow unimpeded movement of the folding-stock mechanism.

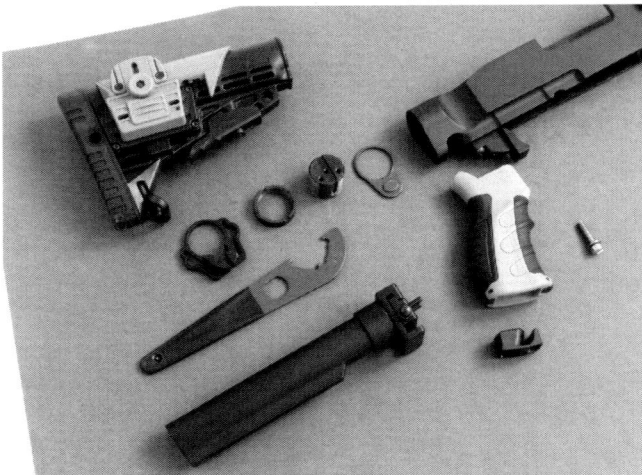

The components required to complete the folding stock and pistol grip assembly of the Evolution Ultimate Match Rifle.

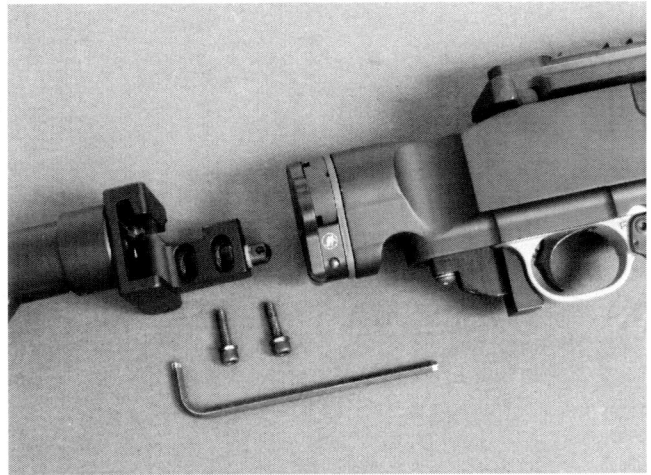

View showing the completed folding-stock tubing ready to be assembled to the ACE receiver block assembly.

6. Install the ACE FAL-style skeleton stock by attaching the folding-stock mechanism to the ACE receiver block. Make sure the buttstock is vertically straight in relation to the rear sight and pistol grip before tightening the assembly screws. You can adjust the height of the stock to your liking. In addition, the folding-stock mechanism is reversible, so you can choose on which side of the gun you want the stock folded. On the 10/22, it is best to have the stock folded on the left so it doesn't interfere with the cocking-handle operation and affect ejection of empty shell casings when the gun is being fired with the stock folded.

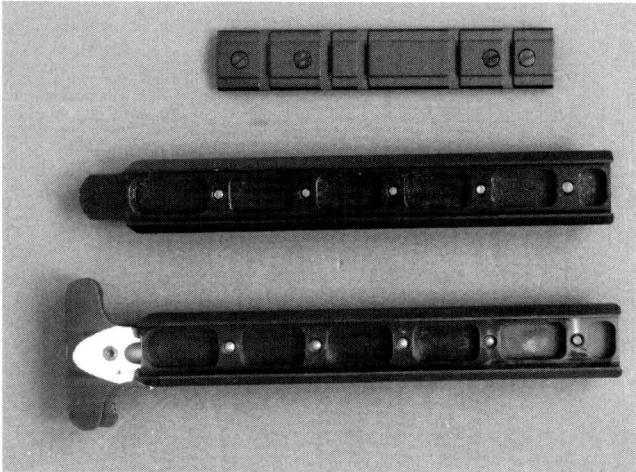

View showing the bottom of the Hahn Precision ;ow-profile riser. The bottom part comes complete with a modified, shortened, nonfunctional AR-15 charging handle assembly. In the center is the original unmodified Hahn Precision accessory. At top is the Ruger scope mount, with which the HP riser will be mated.

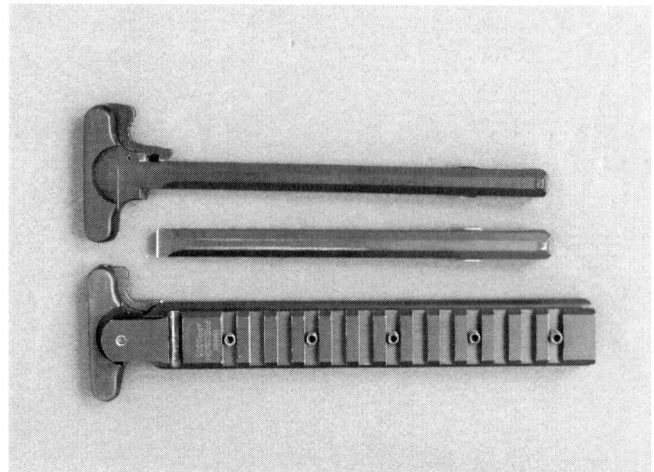

Top view showing the modified (shortened) AR-15 charging handle assembly installed to the HP riser (bottom). The unmodified AR-15 charging handle is seen above; below is the portion of the shortened, modified part installed to the Hahn Precision riser to show the correct cutting point.

7. Install the Magpul MOE pistol grip. (A beavertail grip must be fitted to the radius of the chassis.)

8. Install the factory-supplied scope mount to the 10/22 receiver.

9. Slide the Hahn Precision low-profile riser to the factory scope mount, complete with AR-15 charging handle, until it stops and is seated properly. Tighten the riser's set screws until they are flush to the cross-slots. You will need to trim the bottom of the set screws to the proper length—i.e., until the tops of the set screws level off with the rail's cross-slot. The last screw at the front of the riser will pass the factory scope mount but will still need to be turned down until flush with the cross-slot so it will not interfere with the assembly of accessories. Use this as a spare when needed.

Note: There is a slight looseness in the assembly of the Hahn Precision low-profile riser to the factory scope mount. Normally, the Hahn set screws will tighten up the connection adequately. However, if you prefer a much tighter fit, simply cut a strip of corrugated plastic packaging (normal thickness is about .020 to .022 inch) about 4 1/2 inches long by .174 wide. (This plastic is used on pop bottles, grocery packaging, and firearm accessories packaging.) Place the strips on both sides of the factory rail (below the rail itself) and use cyanoacrylate (e.g., Krazy Glue or Super Glue) to mount the plastic strip underneath the factory rail. This tight arrangement pulls the Hahn accessory down so that it seats flush to the top of the receiver, thereby creating a seamless fit.

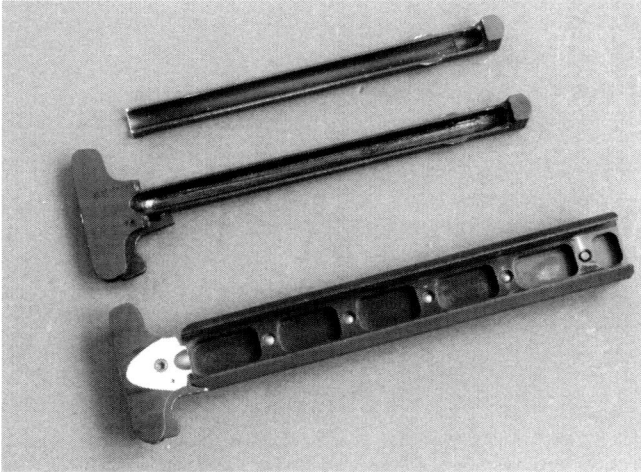

Bottom view of the same parts.

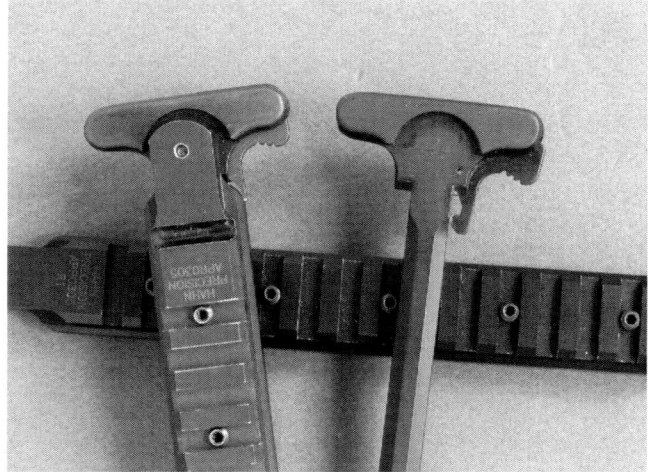

Detail of the modified AR-15 charging handle assembly hook.

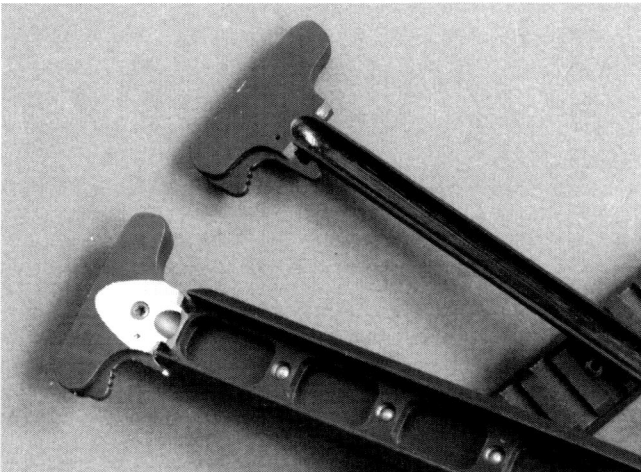

Close-up view of the modified-radius underside section of the AR-15 charging handle assembly. This cutout radius is necessary to clear the rear contour of the receiver and allow the Hahn Precision riser to seat fully forward to connect with the RB scout rail without any gap.

View showing the factory scope mount assembled to the receiver. Note the thin, corrugated plastic sheeting placed on both sides of the rail for tighter assembly with the Hahn Precision riser (shown above). **Note:** The set screws of the riser must be raised to clear the factory scope mount prior to assembly. The bent portion of the plastic sheet is glued at the back of the factory scope mount.

The Hahn Precision riser shown assembled halfway. The plastic sheeting automatically straightens up and tightens the assembly as you push the riser forward.

View showing the Hahn Precision riser fully seated forward and set screws tightened. You may now install your choice of backup iron sights.

The final step for this tactical setup includes the installation of the side-mounted bipod mount and shortened VG1 forward grip. Both accessories are from CAA/EMA Tactical Innovations.

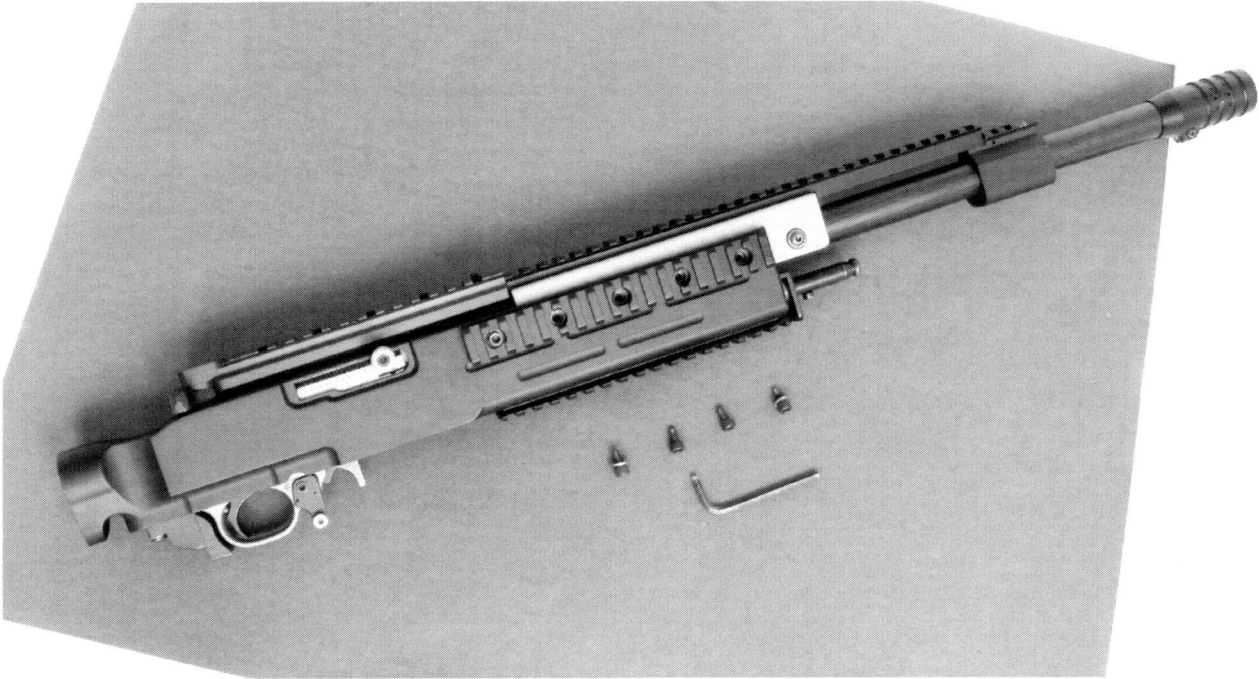

The completed action mated to the Evolution platform. The four side screws secure the matching hand guard.

10. Assemble the action (gun with barrel) to the modular chassis as you would normally do with the regular factory stock and then tighten the stock screw at the bottom of the forearm in the usual manner.

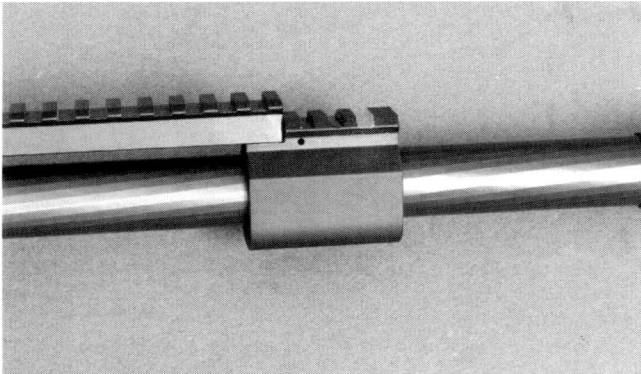

The front end of the RB scout rail must seat straight and level to the squared cut of the gas block, as shown. You will have to cut the required square notch on the gas block for this purpose.

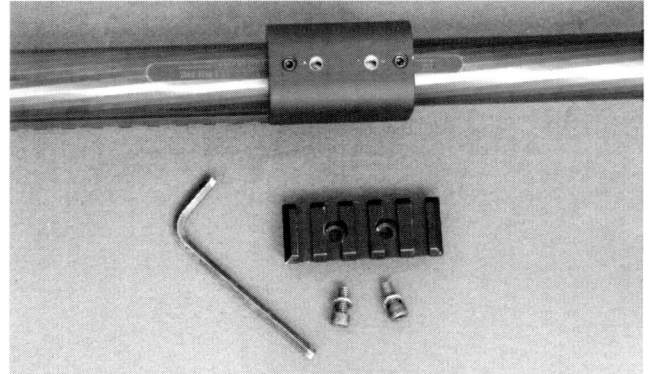

Bottom view showing the Double Star .936 heavy-barrel gas block fully assembled. The double set screws automatically center the part to the squared section where the markings of the barrel were stamped. Installation of the Falcon Industries 2-inch rail below the gas block will require drilling and tapping two holes, as shown.

11. Install the matching top cover to the chassis with the scout rail overhanging on the front. Assemble the side rail screws. (The sling swivel mount, YHM-262, can be readily installed to the threaded hole at the front-end portion of the matching cover, extending over the chassis.)

Photo comparing the 5- and 6-inch YHM rail extensions. The longer rail (middle) has a circular groove at the front cut with a Dremel sander. The front end of the custom Versa-Pod bayonet-style mount will seat on this groove.

11. Roll the pop-can sleeve tightly over the barrel where the gas block will be seated and hold it there firmly. (This part can be improvised by simply cutting a sheet 4 1/2 inches long by 1 1/2 inches wide from a pop can using a pair of scissors.) The sleeve is necessary here to eliminate the looseness of the .008-inch gap in circumference between the outside diameter of the barrel (.920 inch) and inside diameter of the gas block (.936 inch).

12. Slide the Double Star gas block down the barrel until it hits the rim of the aluminum sleeve. At the same time, carefully wiggle and push the gas block against the sleeve until the gas block moves freely rearward over the sleeve with slight hesitation. Keep the pressure (tightness) on the rolled sleeve so it will not move or uncoil itself while you are pushing the gas block in until it is paral-

lel with (covers) the sleeve. Now, line up the front end of the scout rail to the squared notch at the top rear of the gas block and push it in to seat properly before tightening the set screws of the part. At this point, install both backup iron sights to determine if the rear and front sights are perfectly in line. Before proceeding with the alignment adjustment, make sure that the peep sight of the rear sight is set on center. It is best to do this while the gun is resting on a bipod. Make the necessary correction if the sights are not centered properly by simply rotating the gas block on either side until both sights are perfectly in line with each other. Do this before tightening the set screws at the bottom.

13. Install the Falcon Industries 2-inch rail at the bottom of the gas block.

Comparison of the Versa-Pod's Picatinny rail bayonet mount (top left) and the custom-made bayonet-style mount for Evolution chassis (bottom). A partial view of Versa-Pod's Model 1 tactical bipod can be seen on the right.

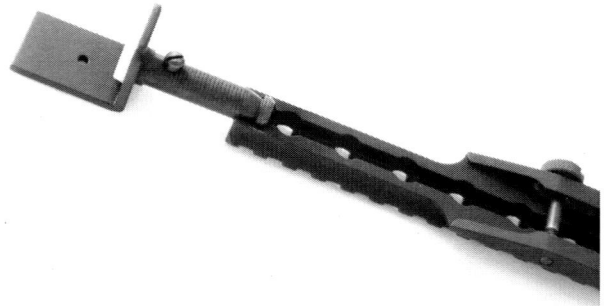

View showing the relationship between the YHM 6-inch rail extension and the custom Versa-Pod bayonet mount in fully assembled mode.

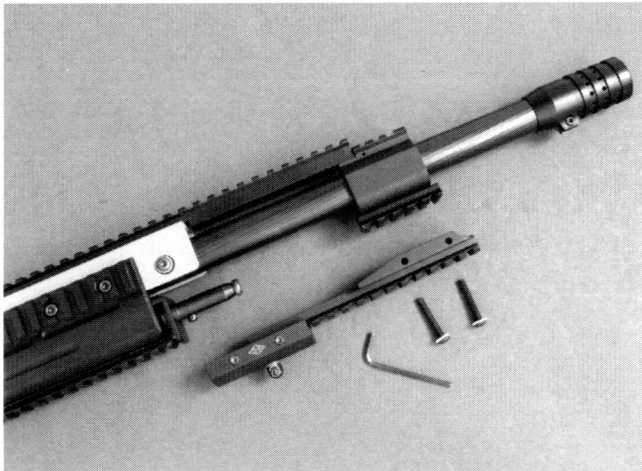

The YHM 6-inch rail extension complete with YHM Harris bipod adapter ready to be installed to the Falcon Industries 2-inch rail.

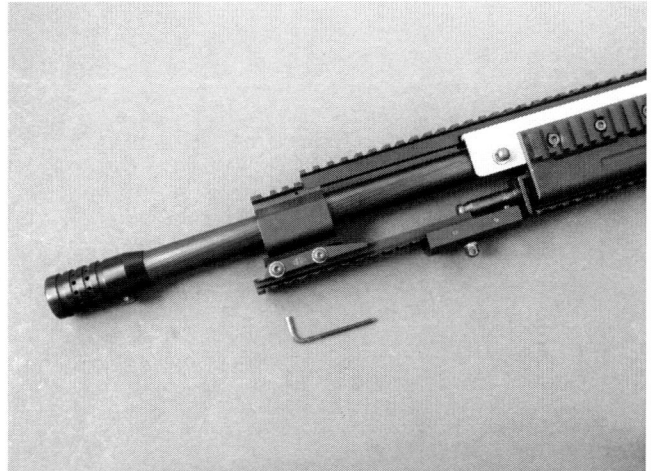

The YHM rail extension fully assembled to the gas block bottom rail as viewed from the left.

14. Attach the YHM 6-inch rail extension to the Falcon rail and push it rearward toward the stock until a portion of the YHM rail front end connects to the tip of the Versa-Pod bayonet-style mount and the assembly screw holes for the YHM bipod adapter line up with the Falcon cross-slots.

View showing the Harris tactical bipod (bottom) ready for installation.

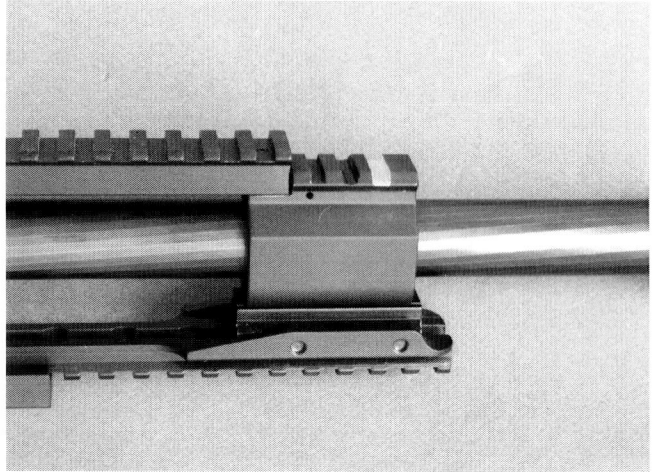

The Double Star cross-slots will need to be cut deeper to accommodate various brands of detachable front sights. Here the front slot has been deepened to fit larger assembly screws of some front-sight brands.

View showing the Harris bipod installed and the Brownells tactical folding front sight assembled to the gas block. The DPMS classic-style front sight, shown above, is used as an alternate accessory when the carry handle–style rear sight is preferred for a vintage look.

View showing the Evolution Ultimate Match Rifle, complete with Harris bipod and NcStar tactical light. The light can be substituted with laser for effective night shooting application.

15. Assemble the YHM Harris bipod adapter to the 6-inch rail close to the chassis. The remaining space at the front of the rail provides ample room for a laser or tactical light.

The Supreme Match Rifle (SMR22) is the final version of the UMR22, featuring Magpul pistol grip complemented by Tapco short vertical grip. The ACE FX folding stock comes complete with CAA cheek riser to match the high-mounted Leapers SWAT long-range scope. Magazines are Ram-Line compact, 40-shot double-stack varieties.

16. Attach the Tapco Fusion short-vertical grip to the forearm rail.

The SMR22 with the stock in folded position. The two magazines seen at the front are Ram-Line ultracompact, 20-shot double-stack magazines. The sling is an M1 .30 carbine issue.

Close-up of the nonfunctional AR-15 charging handle incorporated on the SMR22 model.

17. Attach the Volquartsen stabilization module to the muzzle. If you find the assembly of this part to the barrel loose, you may also apply the pop-can sleeve technique. It works great.
18. Assemble the CAA 30mm scope rings to the Leaper's full-size SWAT scope.
19. Assemble the CAA FAL rifle cheekpiece to the YHM skeleton stock. (**Note:** Install this accessory only when you are using a high-rise scope setup with the scope riding over the rear sight. This cheekpiece is too high when using the standard backup iron sights and must be disassembled from the stock before switching to iron sights.)
20. Assemble the CAA bipod mounts to the side rail of the stock. Install the bipod legs to their corresponding mounts. The legs can be folded on either side of the forearm when not in use, or they can be taken off instantly when needed to lighten the gun when fired from the shoulder.

The gun is now fully assembled and ready to use.

An assortment of parts and accessories that can be alternated for use with the UMR tactical rifle.

Right-side front view of the Evolution Ultimate Match Rifle. The UMR is the first version of the SMR model without the nonfunctional AR-15 charging handle. This gun is dressed with CAA furniture in tan and black. Alternate optic and muzzle devices are also shown with the gun.

Left-side view of the UMR with its stock in folded position. This gun is equipped with Green Mountain 18-inch, fluted heavy barrel and is custom-threaded to accept a variety of AR-15 muzzle devices. It is capable of 1-inch groups at 50 yards.

The SMR-M2 is a "sniper"-type simulator rifle complete with a fake suppressor made by Spike's Tactical. This superb simulator is dressed in Magpul olive drab garb. The SWAT full-size scope is complemented by RRA's 30mm high-rise cantilever scope mount.

Left-rear perspective of the SMR-M2. No matter where you look at this marvel, it has the look and feel of a world-class, rimfire, long-range sniper rifle simulator. The Evolution platform never looked this good, and the SMR-M2 has tack-driving accuracy to back up its looks.

The SMR-M2 with its stock in folded position.

The SMR-M2 setup will accommodate most brands of popular red-dot and full-size scopes shown with the gun.

The SMR-M2 in competition format. The fake suppressor is replaced with a DPMS Levang muzzle brake device. The optic is the battle-proven EOTech 512, complemented by a DPMS tactical rear sight and Brownells AR-15 folding front sight.

Left-side view of the DPMS gas block and YHM short rail extension assembly. The gap between the gas block and the Evolution hand guard is nicely camouflaged with a nonfunctional gas tube for a seamless look. The front end of the non-functional part is assembled directly to the block's gas tubing assembly hole. This is a very simple but sophisticated setup. The only additional work required is making the fake gas tube.

DISASSEMBLY

For a normal disassembly procedure (e.g., for cleaning or maintenance), only the eight side rail screws that retain the matching top cover need to be disassembled along with the stock screw. Lift the top cover and remove it from the gun before separating the action from the stock. The trigger group and bolt assembly can now be accessed for regular maintenance or repair in the usual manner. Reverse this procedure for reassembly.

CHARGING HANDLE MODIFICATION

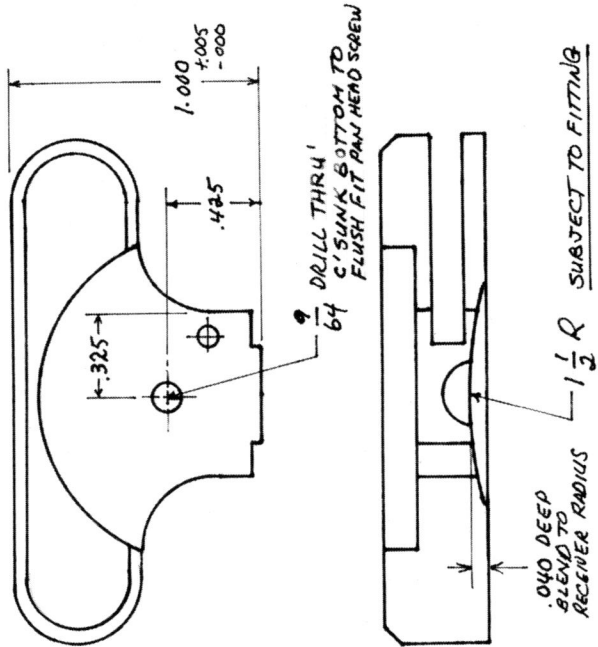

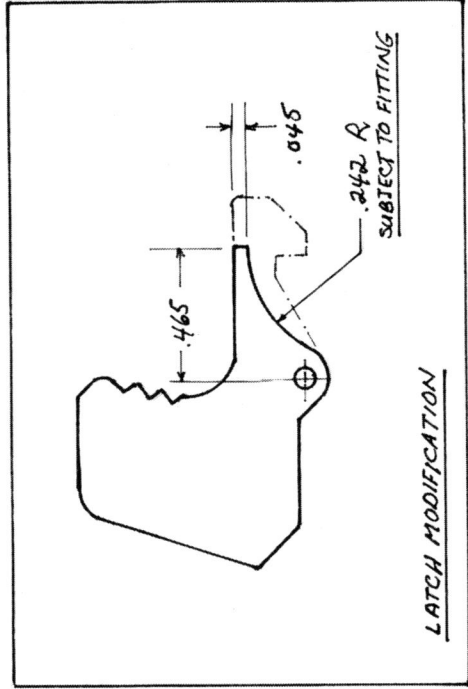

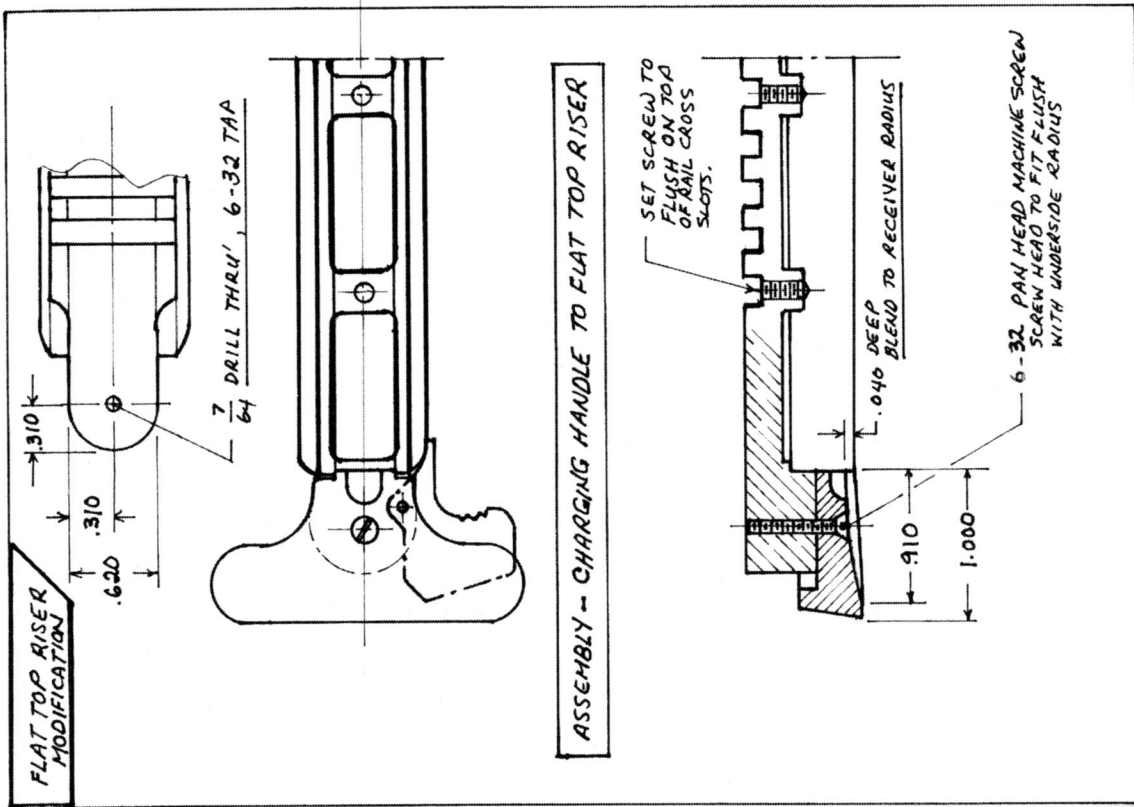

1.000 +.005 -.000

.425

.325

9/64 DRILL THRU' C'SUNK BOTTOM TO FLUSH FIT PAN HEAD SCREW

1 1/2 R

.040 DEEP BLEND TO RECEIVER RADIUS

SUBJECT TO FITTING

.242 R SUBJECT TO FITTING

.045

.465

LATCH MODIFICATION

FLAT TOP RISER MODIFICATION

.310

.310

.620

7/64 DRILL THRU', 6-32 TAP

ASSEMBLY ~ CHARGING HANDLE TO FLAT TOP RISER

SET SCREW TO FLUSH ON TOP OF RAIL CROSS SLOTS.

.040 DEEP BLEND TO RECEIVER RADIUS

6-32 PAN HEAD MACHINE SCREW SCREW HEAD TO FIT FLUSH WITH UNDERSIDE RADIUS

.910

1.000

ASSEMBLY - BOTTOM RAIL & VERSA-POD MOUNT

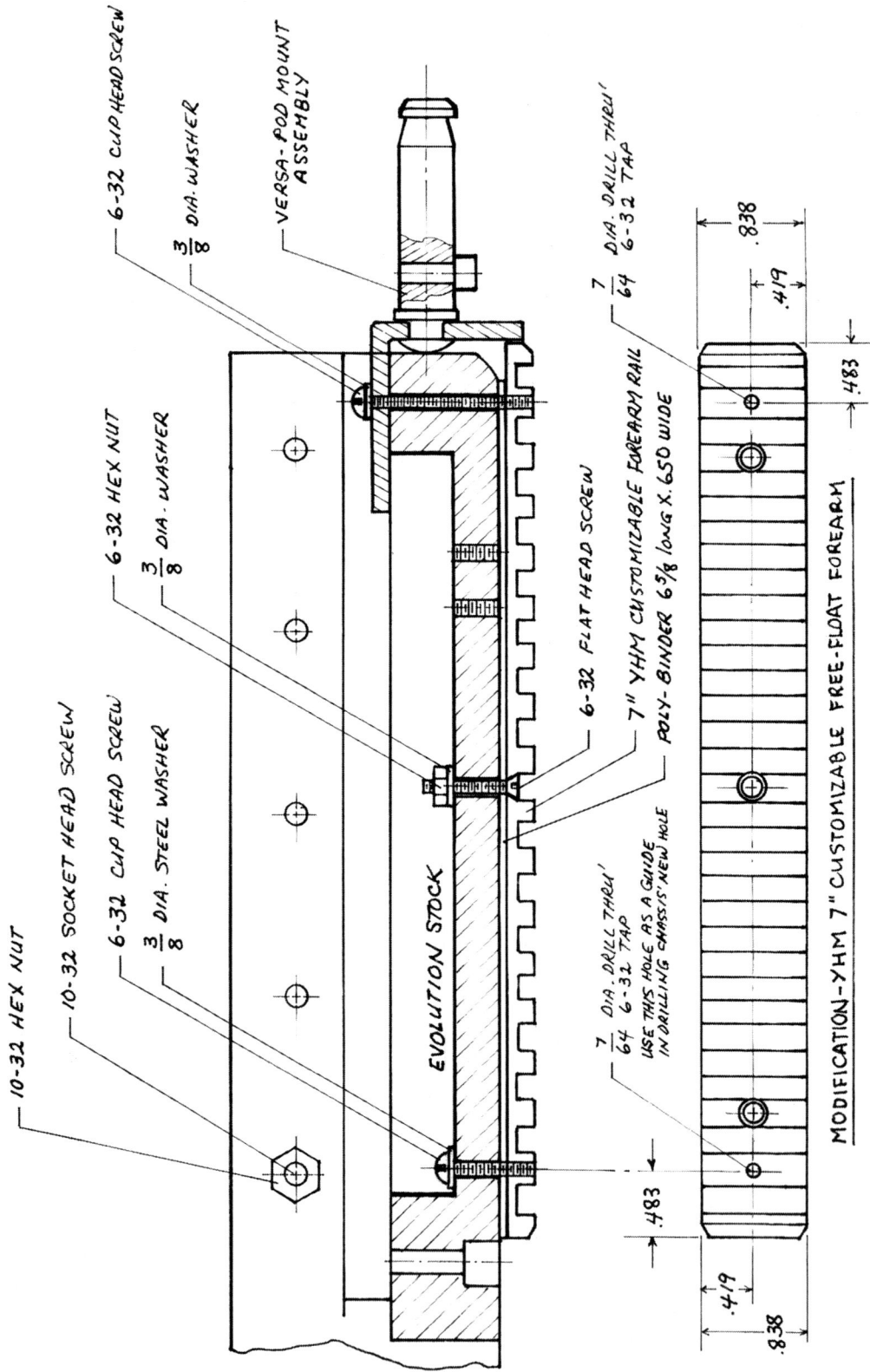

10-32 HEX NUT

10-32 SOCKET HEAD SCREW

6-32 CUP HEAD SCREW

$\frac{3}{8}$ DIA. STEEL WASHER

6-32 CUP HEAD SCREW

$\frac{3}{8}$ DIA. WASHER

VERSA-POD MOUNT ASSEMBLY

6-32 HEX NUT

$\frac{3}{8}$ DIA. WASHER

6-32 FLAT HEAD SCREW

EVOLUTION STOCK

7" YHM CUSTOMIZABLE FOREARM RAIL

POLY-BINDER 6⅝ long X .650 WIDE

$\frac{7}{64}$ DIA. DRILL THRU'
6-32 TAP

$\frac{7}{64}$ DIA. DRILL THRU'
6-32 TAP
USE THIS HOLE AS A GUIDE
IN DRILLING CHASSIS' NEW HOLE

.838

.419

.483

.483

.419

.838

MODIFICATION-YHM 7" CUSTOMIZABLE FREE-FLOAT FOREARM

BUILDING THE RB SUPREME MATCH RIFLE (SMR22)

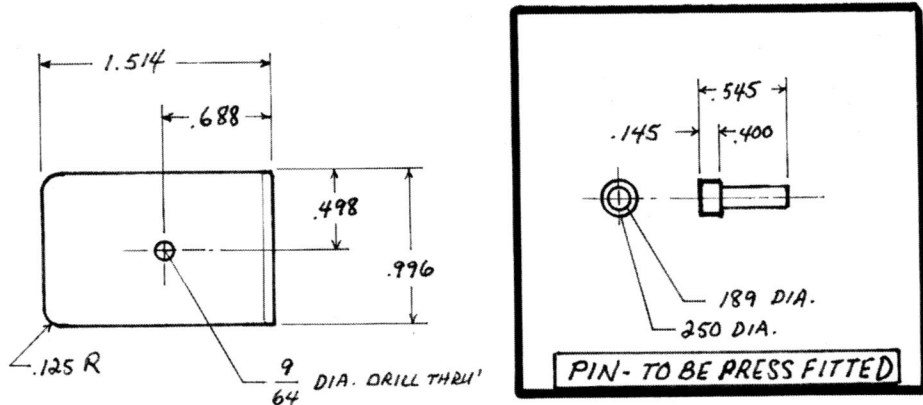

1.514

.688

.498

.996

.125 R

$\frac{9}{64}$ DIA. DRILL THRU'

.145

.545

.400

189 DIA.

250 DIA.

PIN - TO BE PRESS FITTED

BAYONET MOUNT

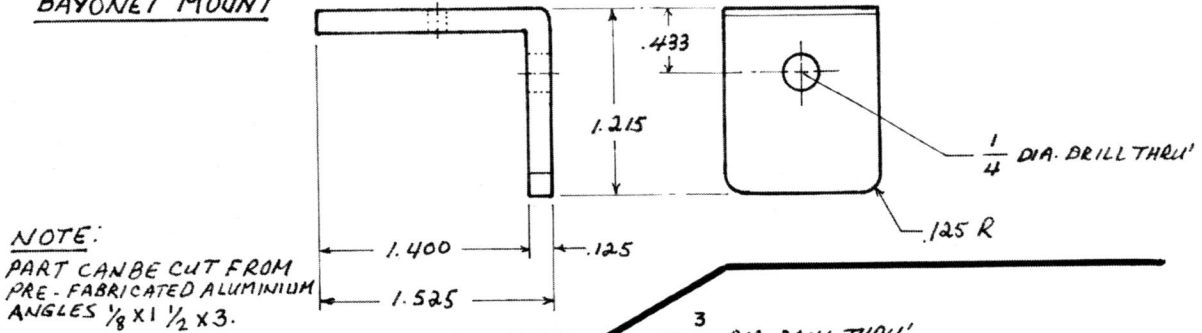

.433

1.215

$\frac{1}{4}$ DIA. DRILL THRU'

1.400

.125

.125 R

1.525

NOTE:
PART CAN BE CUT FROM
PRE - FABRICATED ALUMINIUM
ANGLES $\frac{1}{8}$ X 1 $\frac{1}{2}$ X 3.

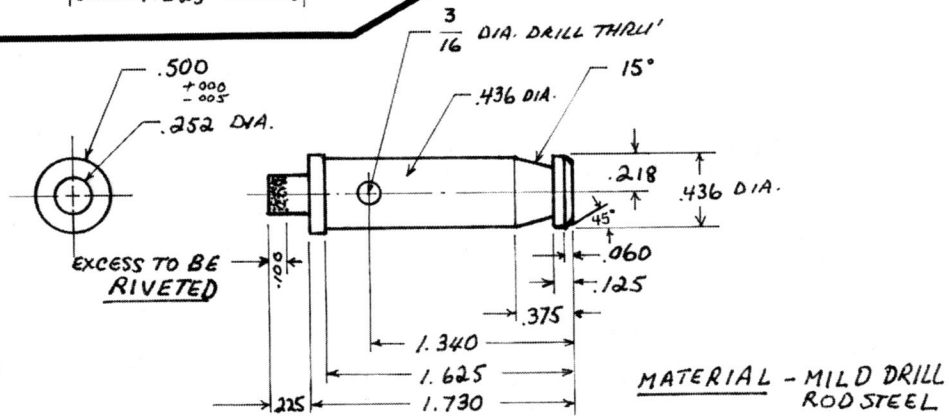

.500
+.000
-.005

.252 DIA.

$\frac{3}{16}$ DIA. DRILL THRU'

15°

.436 DIA.

.218

.436 DIA.

45°

.060

.125

EXCESS TO BE
RIVETED

.375

1.340

1.625

.225

1.730

BAYONET

MATERIAL - MILD DRILL
ROD STEEL

RIVET HEAD

.160

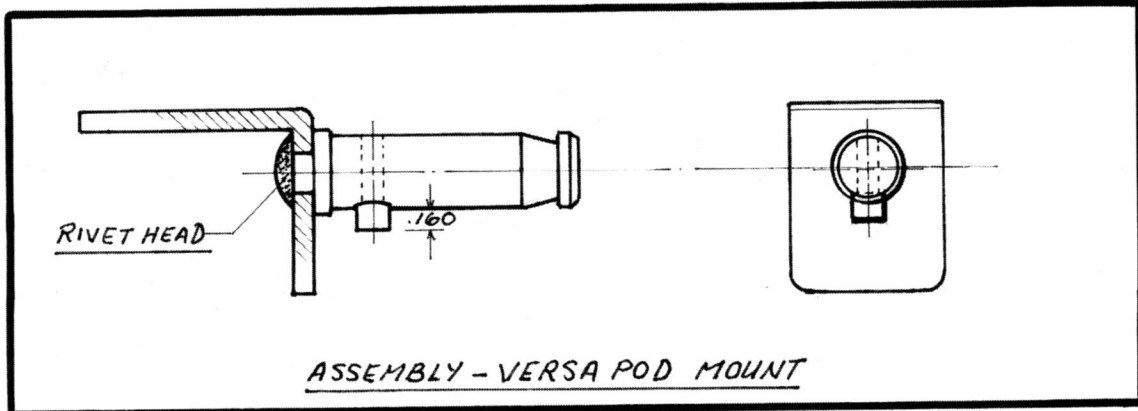

ASSEMBLY - VERSA POD MOUNT

77

CHAPTER 12

Building the Nordic Precision Sniper Rifle Simulator (PSR22)

In the exciting world of modular tactical 10/22 platforms, the Nordic AR-22 easily claims the top spot. This radical AR rifle–inspired platform is a joint effort by two very talented Finnish entrepreneurs, Jarmo Kumpula and Timothy Ubl, the founders of the Nordic Components company. Nordic Components started as a precision manufacturing center, producing parts for the medical industry. The expertise gained from this highly specialized field made Nordic Components the ideal choice for high-precision quality components for the communication, transportation, and defense sectors.

As the company's success in precision aluminum-machining rapidly expanded in the last decade, Kumpula and Ubl (both avid shooters) soon focused their attention on designing and eventually producing specialized products for the shooting industry. Their first effort in this new venture was the AR-22, a custom chassis system primarily designed for tactical upgrading of the very popular Ruger 10/22 carbine. The AR-22 was soon followed by tactical shotgun accessories and AR-15 components. Just recently, the company introduced a complete AR-22 stock kit, which consists of the AR-22 platform, the company's own one-piece aluminum hand guard (mid-length and rifle-length sizes), collapsible M4-style buttstock assembly, and A2-style pistol grip with trigger-guard-gap filler. These components are the same ones used on the Ruger SR22 tactical carbine, with a slight variation on the hand guard vent holes.

Building the Precision Sniper Rifle Simulator (PSRS22) is an exciting experience. It's a wish come

true for those who of us love to build our own tactical rimfire guns utilizing these proven platforms with the always reliable action of the 10/22. Kumpula and Ubl certainly deserve special recognition for designing and producing possibly the finest dress-up platform ever made for the 10/22.

Ruger definitely made a wise decision in choosing this marvelous chassis system when creating its hot-selling SR22 tactical carbine to compete with the new breed of tactical rimfire powerhouses. What makes this superb chassis system so special is its ability to take virtually all types of AR-rifle furniture and accessories, something not possible with other models whose major components are made of plastic composites or whose dimensions are incompatible with commercial or mil-spec standards.

Heckler & Koch's magnificent precision sniper rifle, the PSG1, inspired the overall design characteristics of the custom-made PSRS22. This German-made sniper rifle needs no introduction in military and elite gun collector circles. It is one of the finest and most sophisticated sniper rifles made in the past century and has set the standard by which other modern semiautomatic sniper rifles are still measured today.

The PSRS22 project featured in this chapter is truly an outstanding rimfire simulator model that can replicate the feel and handling characteristics of a full-bore tactical AR sniper rifle when fully equipped for precision target shooting. It's a perfect trainer for beginner snipers who would like to simulate long-distance shooting at 50 to 100 yards, engaging small targets that can project an image equivalent to a man-sized target at 1,000 yards. This kind of shooting practice is quite challenging, and it truly hones one's skill in precision shooting. Competitions in this class of tactical rifle marksmanship are now very popular in the rimfire class. Custom tactical models are pitted against heavy-barreled .22 actions with Barracuda-style skeletonized laminated stocks and ultrasophisticated trigger lock work.

In the early part of 2010, the AR-22 chassis got a face-lift and minor changes to its original format, starting with the horizontal grooves on the side of the main (lower) chassis. The new version has a border-type groove pattern with a large Nordic Components logo just below the magazine ejection port. There are also minor changes in the interior machining to accommodate the 10/22 receiver. In the old chassis, the internal cutout prevented the receiver from moving slightly fore and aft, which meant that the action could not be lifted from the chassis without first removing the upper half from the receiver. The upper half was secured to the top of the receiver by four small screws, which were not designed to be removed on a regular basis during routine field-stripping without the risk of stripping the thread of the aluminum material used in the receiver manufacture. In the new AR-22 chassis, there is a slight gap at the front of the upper and lower halves that allows the receiver to be moved forward just enough for the rear of the trigger guard that locks with the chassis to clear its engagement, thus permitting the action to be lifted up and away from the main chassis.

However, this slight clearance proved insufficient on some of my 10/22s. Only three out of my five guns allowed the disassembly of the action from the lower chassis without first removing the upper half of the chassis from the receiver. Also noticeable was that the baked-on paint finish used on current production 10/22s made mating the upper half of the chassis on some guns difficult. It appears that the AR-22 dimensions were scaled from 10/22s with the original anodized finish that does not add exterior thickness on the surface of the receiver.

Another modification is that the nylon set screw found at the rear of the original chassis has been eliminated, as have the two smaller ones forward of the forearm adapter screws. These three nylon screws were found to be unnecessary in the overall retention of the receiver inside the chassis. The two nylon screws originally found at the front end of the adaptor were omitted in the new chassis. These nylon set screws actually created a serious problem in the original setup because they limited the depth of the threads for the adaptor-retaining screws that connect the receiver to the main chassis. On my AR-22 chassis assembly holes, it didn't take long for these threaded holes to get stripped of their thread after repeated assembly and disassembly of the forearm adapter during regular maintenance. I had the stripped threads rethreaded for a larger hex-type screw, returning the chassis to operation.

It appears that the manufacturer noted this problem and corrected it with a new model that is equipped with a much larger hex screw, one that is the same size as the grip screw. There is also a noticeable lightening of the upper half, which is quite large and heavy for the size of the screws that keep it in place. The unsightly gap between the lower and upper halves still remains on the new model. The problem is only aesthetic, but it can affect the overall appeal of the gun for those who demand perfection. I hope that the manufacturer will introduce a third version of the AR-22 platform with a seamless fit at a later date. For now, the first- and second-generation AR-22 chassis will do the job just fine. Just be careful not to overtighten the forearm adapter screws of the Model 1 chassis—it will definitely strip the threads since there is barely 1/8 inch of screw holding the adaptor to the chassis.

Among the three custom platforms featured in this book, the AR-22 has the best engineering and design characteristics. Its receiver contours at the rear for the buttstock extension, and the grip-mounting base is perfectly machined in very close proximity to the AR rifle lower. As a result, custom grips (such as the CAA and Magpul with beavertail grip extension) simply drop into place with little or no fitting required. On the original AR-22 chassis, the top front end of the grip that butts against the rear upper section of the trigger guard needed some inletting to clear it. This is no longer required on the new chassis since the receiver can be moved slightly forward to minimize contact, thus allowing the action to seat properly inside the chassis.

Another excellent design characteristic of the AR-22 is the forearm adaptor. The threaded forward section of the adaptor that accepts the forearm is also scaled from the AR receiver dimension, allowing both two-piece and one-piece free-float forearms to be installed on it, something not possible with the other kits. These excellent design features make the AR-22 the ideal chassis for AR fans who want to build a world-class tactical rimfire weapon system that can replicate the versatility and aesthetics of the finest high-end full-bore AR competition/sniper rifles, carbines, SBRs, and PDWs.

The procedure for building a sophisticated PSG1 sniper-style .22-caliber rifle out of a 10/22 with an AR-22 chassis is straightforward and simple to follow. However, it does require that the furniture and accessory combinations coordinate so that the end result is not only attractive but also user friendly and ergonomically perfect. Although there are a number of models presented using the same chassis and barreled action, their furniture, pistols grips, forearms, muzzle devices, rails, and so forth are different. This gives you an idea of how the guns look in various color combinations with different accessories.

In building the AR-22 models, I used the ACE folding stock because it is as solid and strong as a fixed stock, yet it makes a full-size gun more compact for storage and transport. The installation of this well-made folding stock to any regular-production gun will improve its aesthetic appeal and overall value. Refer to the accompanying photo sequences for installation procedures of folding stocks, pistol grips and forearms (one-piece and two-piece types), rails, optics, and so forth on the AR-22 chassis.

Other related accessories that you may want to add or delete to these individual setups to suit your particular needs, tastes, or requirements are pretty straightforward to accomplish. Just use your skill and imagination. If you are building the gun for yourself, assemble one that suits and pleases you. If you are doing it for somebody else, make sure that you take into account their individual needs or tastes.

The PSRS22 model incorporates the following parts and accessories:

1. Nordic AR-22 chassis (old or new type)
3. DPMS Panther Tactical Palm Grip
4. CAA VG1 forward grip (shortened to three-finger groove pattern)
5. CAA BSR 8–12-inch side-mounted bipod
6. ACE ARRB (receiver block assembly)
7. ACE PN (CAR stock block)
8. ACE FSM1 (folding-stock mechanism)
9. Rock River Arms (RRA) entry buttstock (available in black or green)
10. Midwest Industries (MI) raised gas block for .0750 OD barrels
11. MI sling adapter (loop or slot, your choice)
12. YHM-9636 lightweight free-float four-rail forearm (rifle length)
13. YHM-9896 (bull barrel, comes threaded for AR muzzle accessories). A section of the front end must be turned to .0750 inch, measured from the tip of the forearm (assembled to gun) to the threaded portion of the muzzle to accommodate the inside diameter of the MI gas block.
14. YHM-9484-B forearm endcap for bull barrel
15. YHM-9680 flip rear sight
16. YHM-9584 gas-block-mounted front flip sight (alternate: Brownells CAR-15 flip-up tactical front sight)
17. YHM-9469 rotating swivel stud
18. Muzzle devices: FS fake suppressor modified to FG-42 style muzzle brake. Alternative muzzle accessories: Tactical Innovations or Spike's Tactical fake suppressors. The flash hider is a Vltor VC-1.
19. Brownells AR-15 flattop riser or Leupold Mk-1 integral mount
20. Optics: UTG Leapers 30mm SWAT 3-12 x 44 full-size AO mil-dot RGB EZ-Tap scope
21. Scope rings: CAA 30mm (CSR3)
22. Magazines: The AR-22 chassis will accommodate all types of 10/22 magazines.
23. Tactical light/laser combination: TLR-2 or SIG SAUER STL-900L (**Note:** The Chinese-made UTG laser as seen on the model guns did not meet my expectations and therefore cannot be recommended here. It was used here for photo representation only. It is of poor quality and difficult to adjust for both windage and elevation to hit the point of aim at any given distance.)

PUTTING IT TOGETHER

The Nordic AR-22 is the most versatile among the three 10/22 custom chassis selected, though all three take virtually any furniture and pistol grips available for AR-15/M16 rifles. The superb PSRS system is put together in the following order.

The correct arrangement of the ACE receiver block assembly and CAA one-point sling mount prior to assembly on the Nordic AR-22 chassis. You will need an M4 stock castle nut wrench (above) for this job.

View showing the ACE receiver block installed to the AR-22 chassis. Remember to leave enough space for the circular ring of the sling loop adapter over the castle nut.

1. Assemble the ACE ARRB components to the rear of the AR-22 chassis, leaving at least .135 inch of the ARRB protruding above the castle nut to accommodate the Midwest Industries sling adapter. You will need an M4 castle nut wrench for this procedure.

View showing the Midwest Industries sling adapter installed. Note the slight protrusion (.030–.035 inch) of the receiver block over the sling adapter to allow unimpeded movement of the ACE folding-stock mechanism when activated.

2. Install the Midwest Industries sling adapter over the castle nut and tighten the assembly screw. There must be at least a .035-inch protrusion of the ARRB above the sling adapter to allow free movement of the folding-stock mechanism when operated.

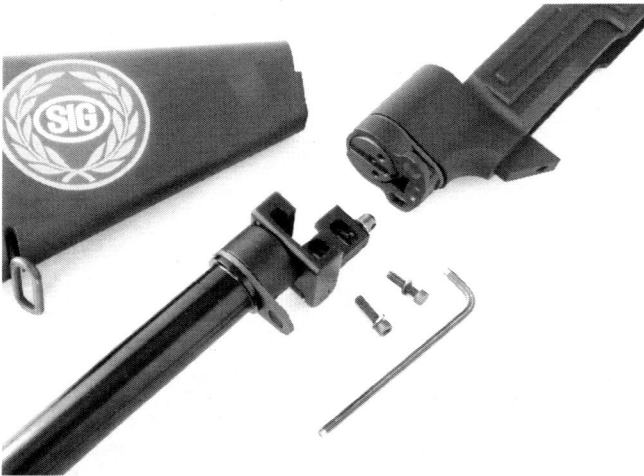

View showing the fully assembled ACE CAR stock block and RRA stock extension tube to the folding stock mechanism (bottom left). The folding stock assembly is now ready to be installed to the AR-22 chassis using a hex wrench. (**Note:** Make sure that the receiver end plate is perfectly centered vertically at the bottom prior to tightening the stock extension to ensure that the plastic stock is positioned correctly during the final assembly.)

3. Install the ACE PN (CAR stock block) to the folding-stock mechanism (ACE FSM).

View showing the folding stock (minus the RRA entry-sized plastic stock) installed on the AR-22 chassis. To complete the assembly, simply slide the plastic stock to the matching receiver extension tube and tighten the assembly screw at the rear of the butt with a large screwdriver.

4. Install the RRA tactical stock tube assembly to the ACE PN. Make sure the stock spacer is vertically straight in the downward position before tightening the stock tubing. The M4 stock wrench has a square slot at the end of the handle and is designed for this purpose.

View showing the compact entry folding stock assembly installed in extended position. The AR-22 chassis is now ready for target-style pistol grip installation. Shown are the DPMS Tactical Palm Grip (right) and the CAA UPG16 with target stand.

5. Install the entry-sized plastic buttstock to the tubing and tighten the assembly screw at the rear top section of the butt using a large screwdriver.

Next to be assembled after the pistol grip installation is the AR-22 upper to the top of the 10/22 receiver using the supplied screws.

6. Install the buttstock assembly to the ACE ARRB via the folding-stock mechanism (ACE FSM).

7. Install the DPMS Panther Tactical Palm Grip to the lower chassis. The CAA UPG16 with target stand can be used as an alternate grip for this setup if preferred.

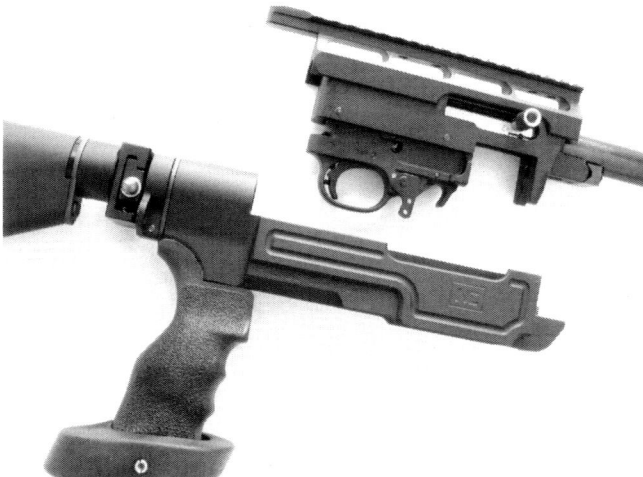

Assemble the 10/22 action to the AR-22 chassis as you would normally do with the factory stock. (**Note:** Not all 10/22 actions will assemble freely with the AR-22 chassis. If this problem is encountered, assemble the upper chassis after the gun's action is fully seated to the lower chassis.)

8. Assemble the AR-22 upper to the 10/22 receiver using the four screws supplied. If the receiver will not seat freely to the lower chassis with the AR-22 upper installed to the top of the receiver during assembly, remove the upper and reinstall after the receiver is fully seated to the chassis.

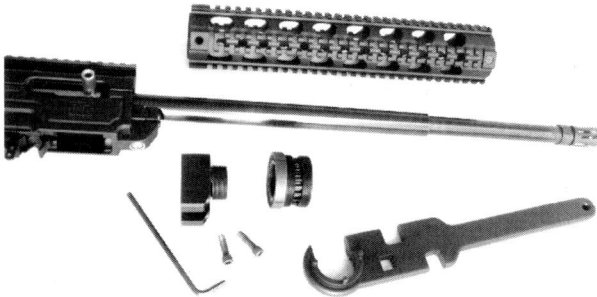

After the assembly of the upper and lower chassis of the AR-22 to the 10/22 action, the next step is the installation of the forearm group to the AR-22 adaptor. This will require an armorer's wrench (Smith Enterprise wrench shown) or YHM forearm wrench, available from YHM.

View showing how the YHM barrel nut is assembled to the AR-22 forearm adaptor using the Smith Enterprise tool. YHM's forearm wrench tightens both the barrel nut and matching jam nut. Use the YHM tool for this application to get the best result.

9. Install and tighten the supplied stock screw to secure the receiver to the chassis.

   This sequence completes the assembly of the buttstock and receiver group to the chassis. The next step is the installation of the forearm, gas block, and muzzle device. For this step, follow the sequences outlined above.

10. Install the barrel nut (YHM-9483) to the AR-22 forearm adaptor. You will need the YHM-9621 forearm wrench to precisely align the barrel nut and tighten the jam nut after the forearm rail is lined up with the AR-22 rail.

For correct sequencing, assemble the jam nut first to the adaptor, followed by the four-rail forearm. Before tightening the jam nut after the forearm is fully seated, make sure that the top rail of the forearm is perfectly aligned to the AR-22's top rail. Next, assemble the anti-rotation screws to prevent any movement of the forearm after the jam nut is tightened.

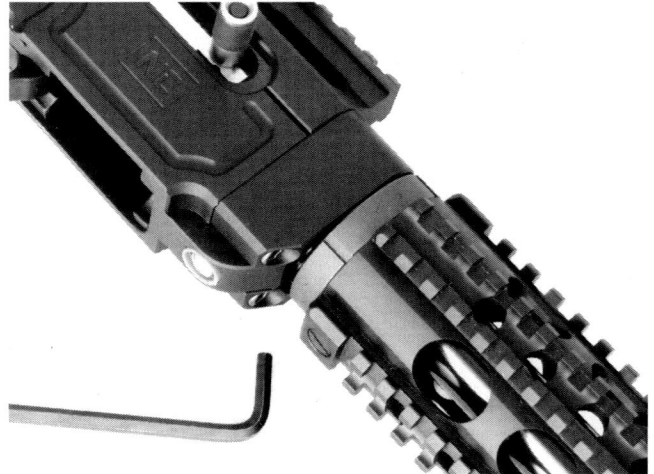

Install the forearm adaptor complete with forearm assembly to the AR-22 lower chassis. Two large hex screws secure the forearm adaptor in place. (*Caution*: Do not overtighten the assembly screws, as you may strip the threads.)

11. Install the jam nut (YHM-9433) to the barrel nut all the way to the rear so it's touching the adaptor. Install the forearm (YHM-9636 rifle length) to the barrel nut as far as it will go and line up any of its four-sided rails with the AR-22 rail. Tighten the jam nut against the forearm using the YHM combination wrench. Make sure that the grooves of the barrel nut are in line with the anti-rotation screws (YHM-9434) before installing and tightening these retaining screws so you don't damage the nut.

Assemble the matching end cap as shown (optional). Next, install the MI raised-rail gas block (with fake piston and spring) and then finish off with Vltor VC-1 compensator installation.

12. Install the forearm endcap (YHM-9484-B) to the front end of the tubing (optional).

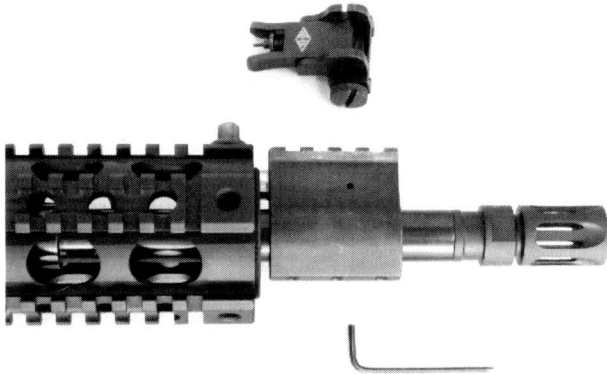

Install the YHM gas block–mounted folding front sight to the MI gas block. Note the fake piston partly showing between the forearm end cap and gas block.

View showing the YHM folding front sight installed and in deployed position. The MI gas block utilizes three set screws, which must be tightened only after the front sight is perfectly lined up with the matching rear sight. It is best to do this during an actual shooting session.

13. Install the AR-22 forearm adaptor complete with YHM forearm group to the lower chassis. The turned section at the front of the barrel (.750 OD) must be parallel to the inside rim of the endcap. There should be enough clearance for the barrel to pass through the endcap without touching it.

14. Assemble the MI gas block to the barrel in reverse order (large hole facing the forearm) and stop approximately 1 inch before the part hits the rim of the forearm. This will provide enough space for inserting the fake gas piston assembly to the gas block.

The bottom rail of the YHM rifle length forearm can accommodate forward grip, light/laser, or bottom mounted bipod all together. The bipod shown above is a side-mounted type with detachable legs made by CAA/EMA Tactical.

15. Install the fake gas piston and spring to the MI gas block large hole and then push the gas block forward, aligning the smaller-diameter front end of the fake gas piston to the endcap hole. Leave a gap of at least .125 inch between the endcap and gas block to expose a portion of the fake piston for the desired exterior effect. (See workman's drawing on page 92.)

16. Line up the gas block rail with the forearm rail before tightening the set screws.

17. Install the muzzle device (fake suppressor preferred for this setup) if the muzzle is threaded for AR-15 rifle accessories.

18. Install the CAA BSR bipod side mount to the YHM forearm side rails.

The shortened CAA VG1 grip, TLR-1 light, UTG laser (on top rail), and CAA side-mounted bipod installed to the forearm rails, with the right leg detached. The versatility of the YHM four-rail forearm is put to a good use in this setup.

19. Install the CAA VG1 forward grip. The grip used in this setup is shortened, reducing the four-finger groove pattern to three for a more streamlined look. I believe that the standard-length grips are too long and bulky for all practical purposes. Shorter forward grips are becoming more the norm and are much preferred by today's professional marksmen.

The final accessories to be assembled to the gun are the matching YHM tactical flip rear sight and Leapers full-size SWAT 30mm tactical scope mounted to Brownells AR-15 flattop riser.

20. Install tactical light in front of the forward grip.
21. Install a separate laser system on the top rail of the forearm if needed. This setup will only work with a folding front sight.
22. Install the backup iron sights at their designated location.
23. Assemble the CAA CSR3 (30mm scope rings) to the Leapers 30mm SWAT scope.

Right-side view of the completed PSRS with alternate grip and muzzle device shown below. The Black Dog 50-shot drum and Ram-Line 40-shot compact banana-type magazines provide exceptional firepower and fun.

Same gun with the stock folded and with the large muzzle brake substituted.

Close-up of the PSRS rifle folding stock, which is a great accessory for making a long-barreled rifle more compact for transport and storage, as well as shooting on assault mode in confined spaces using a laser as an aiming device.

Close-up of the muzzle devices, TLR-1 light, and NcStar laser. The front sight must be in the folded position before using the laser in this setup. Note the YHM sling mount behind the front sight.

24. Assemble the scope to the Brownells AR-15 flattop riser.
25. Assemble the CAA bipod legs to their side mounts in the YHM forearm.
26. Assemble the Brownells flattop riser complete with full-size scope to the AR-22 upper rail. The gun is now ready for action.

Front-left view of the PSRS rifle with Ram-Line compact 40-shot banana magazine ready for action. Note that the front sight is in the folded position, allowing a clear pathway for the laser beam to pass unobstructed.

Right-front view of the PSRS rifle. This superlative tactical 10/22 simulator is the ideal training apparatus for low-budget police departments as well as junior military cadets prior to full-bore weapon exposure.

This concludes the assembly of this marvelous rimfire sniper simulator system, which is ideally suited as an economical trainer for those aspiring to be police or military marksmen. For regular cleaning, repair, or maintenance, follow these simple steps.

STEPS FOR CLEANING, REPAIRING, OR MAINTAINING RIFLE

1. Disassemble the muzzle device from the barrel.
2. Loosen the set screws and disassemble the gas block from the barrel. Be careful not to lose the spring-loaded fake gas piston while easing out the gas block from its seated position.
3. Disassemble the bipod legs from their respective mounts.
4. Disassemble the AR-22 forearm adaptor from the lower chassis in the usual manner. You need not remove the accessories already installed on it.
5. Disassemble the stock screw to release the receiver from the lower chassis.
6. Lift the action of the gun from the lower chassis, front end first. If you cannot take the action out, remove the upper from the top of the receiver by unscrewing the four assembly screws first. Reverse the procedure during reassembly of the action to the lower chassis if you encounter this problem again.
7. The action is now out of the chassis and ready for normal disassembly procedures.
8. Reverse the above sequence for reassembly.

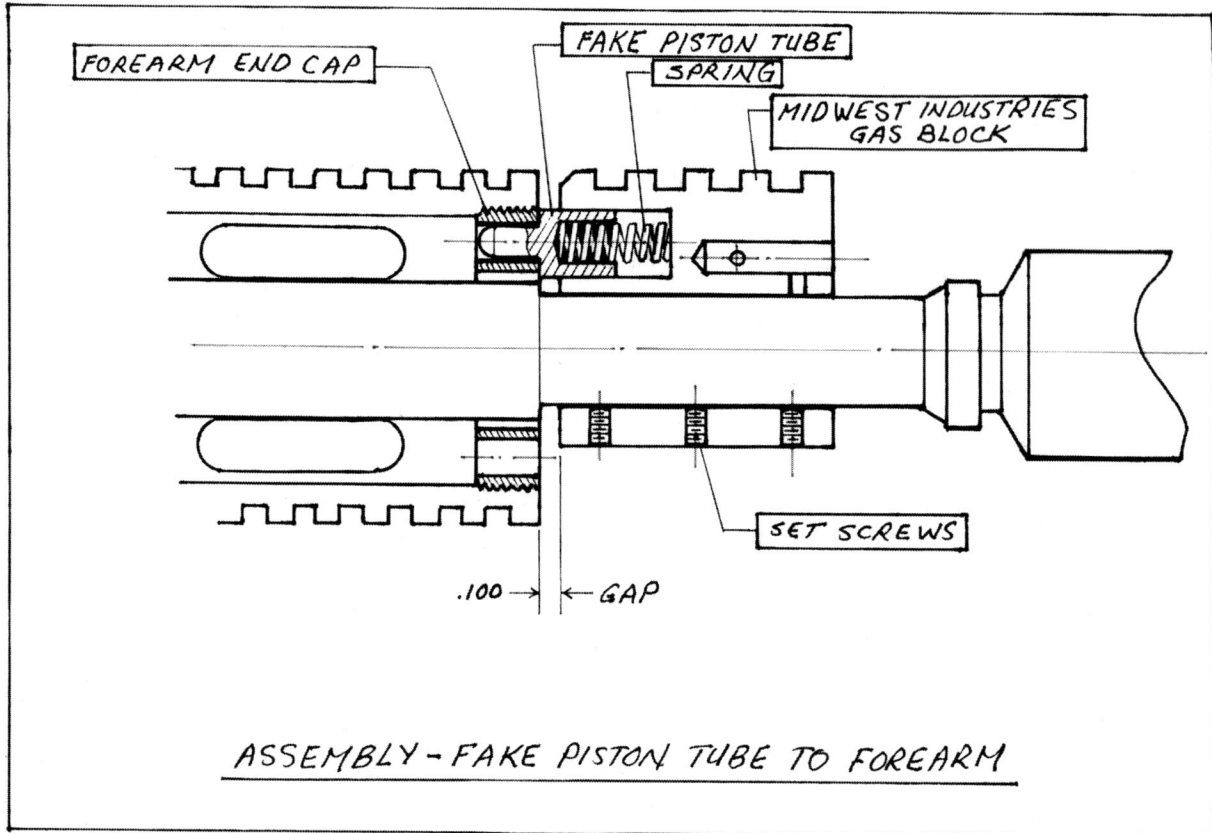

FOREARM END CAP

FAKE PISTON TUBE

SPRING

MIDWEST INDUSTRIES GAS BLOCK

SET SCREWS

.100 → ← GAP

ASSEMBLY - FAKE PISTON TUBE TO FOREARM

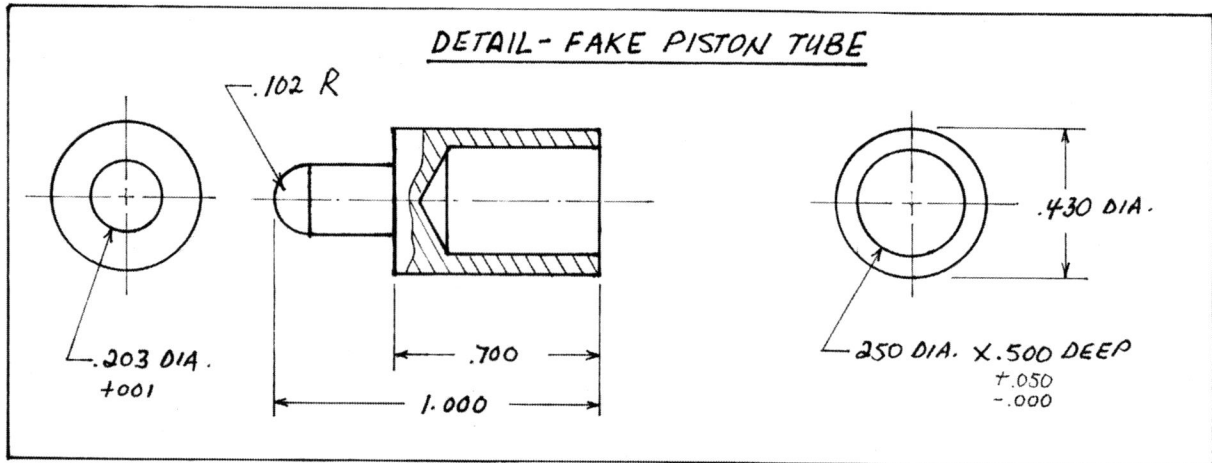

DETAIL - FAKE PISTON TUBE

.102 R

.203 DIA.
+001

.700

1.000

.430 DIA.

250 DIA. X .500 DEEP
+.050
-.000

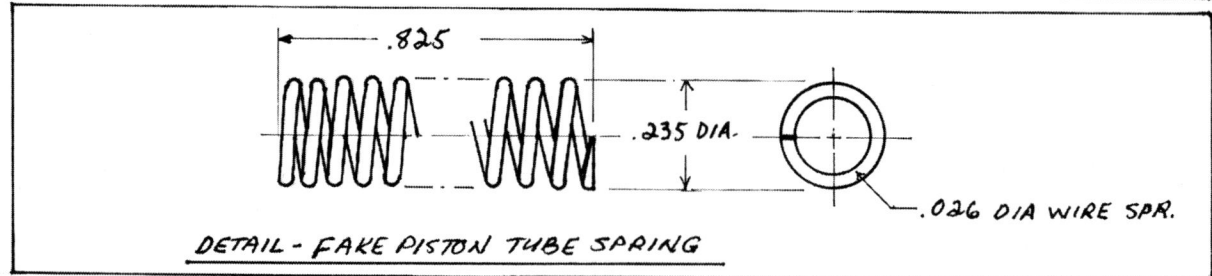

.825

.235 DIA.

.026 DIA WIRE SPR.

DETAIL - FAKE PISTON TUBE SPRING

CHAPTER 13

Building the Fabsports Marksman Challenger Rifle (MCR22)

This limited-edition Canadian-made 10/22 chassis system is definitely one of the best among the modular-type 10/22 platforms that can compete toe-to-toe with the very popular Nordic AR-22. The Fabsports (FS556) has a seamless fit between the lower and upper halves of its clamshell design. Its overall arrangement, however, takes longer to assemble and disassemble during routine cleaning. In the Nordic system, the forearm adaptor is secured to the lower half by two large Allen screws, the same size as the AR-15 grip screws. In the FS design, the forearm adaptor is secured directly to the main chassis by five medium-sized Allen screws. This arrangement is more solid than the AR-22 setup, but the five assembly screws on the FS take longer to disassemble.

Another disadvantage of the FS design is that it requires disassembly of the barrel from the action first, before the receiver can be lifted out of the main chassis. The FS upper is secured directly to the front and rear of the lower half by two medium-sized Allen screws, ensuring a tight fit. The internal cutout of the main chassis allows adjustment for centering the action and barrel to the free-float forearm, something the AR-22 is lacking. A rubber-tip set screw is positioned on the upper half and acts against the top of the receiver during assembly for a tight lockup. The FS system will only accommodate one-piece free-float forearms that are made for AR-15 rifles, as well as pistol grips and buttstocks (fixed M4 style or folding).

As with the Evolution and AR-22 chassis systems, the FS can also be transformed into many different tactical weapon profiles; the number is limited only by the builder's imagination and resources. To build this extraordinary tactical 10/22 competition gun, you will need the parts and accessories listed below. If you are planning to purchase a 10/22 for this project, buy the cheapest 10/22 carbine model you can find. A good used one is ideal here since you need just the complete action of the gun minus the barrel and factory stock.

Note: At the time of this writing, the FS556 chassis system is temporarily out of production. If you are interested in purchasing this SIG-style chassis system, you can contact Fabsports directly in Canada for future availability (www.fabsports.ca).

The Fabsports MCR22 features the following parts and accessories:

1. CAA CBS six-position collapsible stock (green, commercial)
2. CAA ACP adjustable cheekpiece (black)
3. CAA 6–8 inch bipod with Picatinny rail
4. CAA UPG16 (black with green inserts). The beavertail was cut and squared at the top to fit chassis.
5. Magpul AFG angled foregrip (OD green). Alternate: Tapco Intrafuse short vertical grip (OD green).
6. YHM-9637-DX Specter Length free-float forearm
7. Magpul MOE stock cheek riser (available in 1/4-, 1/2-, and 3/4-inch height)

8. YHM-9484-B forearm endcap for bull barrel
9. YHM-9469 threaded sling swivel stud
10. YHM-9388 railed bull barrel gas block (alternate: DPMS .936 4-rail gas block)
11. YHM-220 EOTech rail riser
12. DPMS six-position carbine stock tube (commercial)
13. DPMS detachable tactical rear sight (alternate: YHM-643K solid A2-style rear sight)
14. LMT tactical front sight; to be used in conjunction with YHM-9388 raised gas block only. Alternate: YHM-9627, same-plane, flip front sight or YHM-9830A same-plane, fixed front sight.
15. ACE ARRB AR-15 receiver block assembly
16. ACE PN CAR stock block
17. ACE FSM1 folding-stock mechanism (if you prefer a push-button type, order the ACE FSM-PB)
18. MI AR-15 sling adapter (loop or slot, your choice)
19. Green Mountain 18-inch fluted heavy barrel with optional 1/2 x 28 threads for AR-15 muzzle devices
20. Carlson Mini Compensator
21. Optic/red-dot EOTech or Bushnell Holosight
22. Magazines: the FS556 chassis will accommodate all banana-type magazines but will prevent drum or teardrop-shaped magazines from seating fully unless the magazine chute (skirt) is modified.

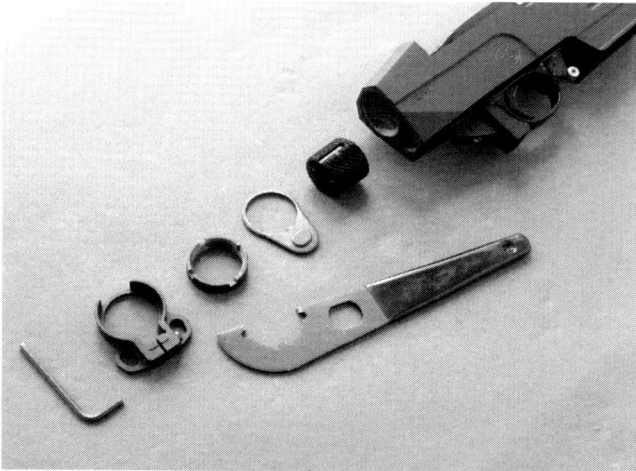

View showing the ACE receiver block assembly and Midwest Industries loop sling adapter ready to be installed to the FS556 chassis. An M4 stock wrench will be needed here.

View showing the ACE receiver block assembly installed to the FS556 chassis. Note the protruding section of the receiver block extending over the castle nut to accommodate the sling-adapter ring.

PUTTING IT TOGETHER

1. Assemble the ACE ARRB to the back of the FS lower chassis, leaving enough room for the installation of the MI sling adapter over the castle nut. You will need an M4 carbine stock castle nut wrench to do this job.

Close-up view showing the MI sling adapter fully installed over the castle nut ring. There must be a slight protrusion of the receiver block over the castle nut to allow unimpeded movement of the folding-stock mechanism.

2. Assemble the MI sling adapter to the castle nut. Make sure there is at least .030 inch of the ACE-ARRB protruding past the sling adapter to allow unimpeded movement of the folding-stock mechanism when operational.

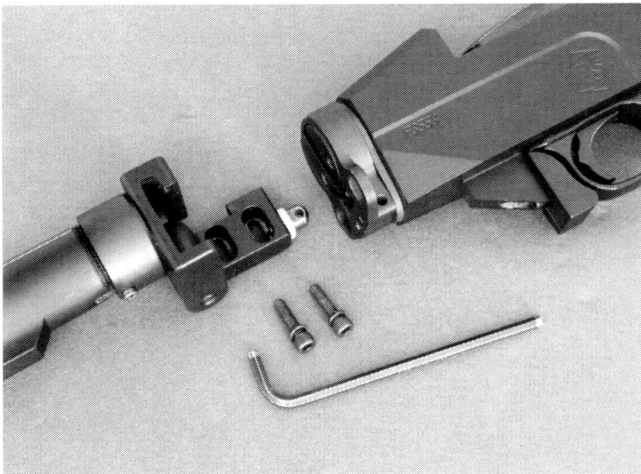

The folding-stock assembly ready to be installed to the ACE receiver block.

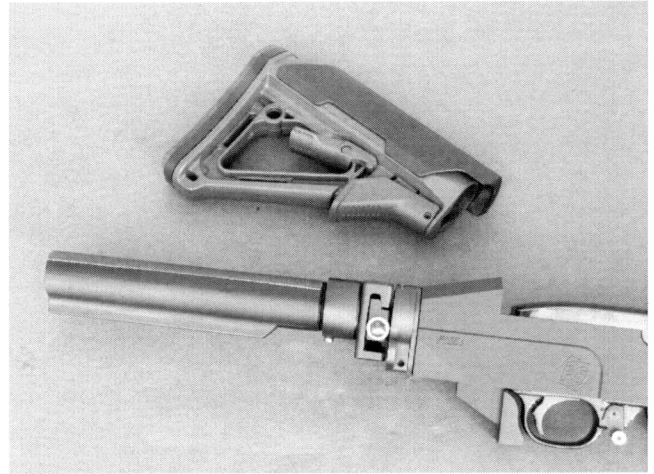

The folding-stock mechanism complete with CAR stock extension assembled to the matching receiver block. Note the slight gap between the MI sling adapter and receiver block, which allows unimpeded movement of the folding-stock mechanism. The Magpul CTR, complete with optional cheek riser, is seen above.

3. Assemble the DPMS six-position carbine stock tube to the ACE PN (CAR stock block) and make sure it is vertically straight before tightening the set screw.

The matching two-tone (green/black) CAA UPG16 (with the beavertail cut off) ready for installation to the FS556 chassis. The Magpul CTR stock in olive drab with black cheek riser nicely complements the CAA pistol grip.

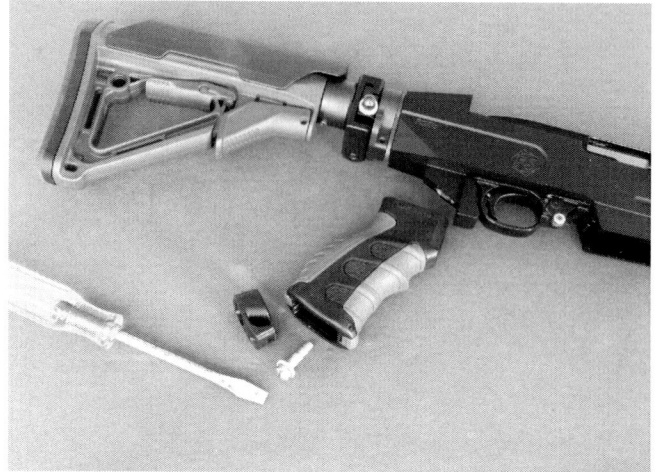

4. Install the CAA CBS stock to the carbine stock tube.
5. Install the CAA ACP to the buttstock.
6. Install the CAA UPG16 (two-tone grip) to the lower chassis.

After completing the assembly of the buttstock group to the FS556 chassis, the next step is to assemble the 10/22 action (minus the barrel) to the FS lower chassis with the front end inserted first as shown. The matching upper is shown above painted in olive to match the two-tone furniture combination.

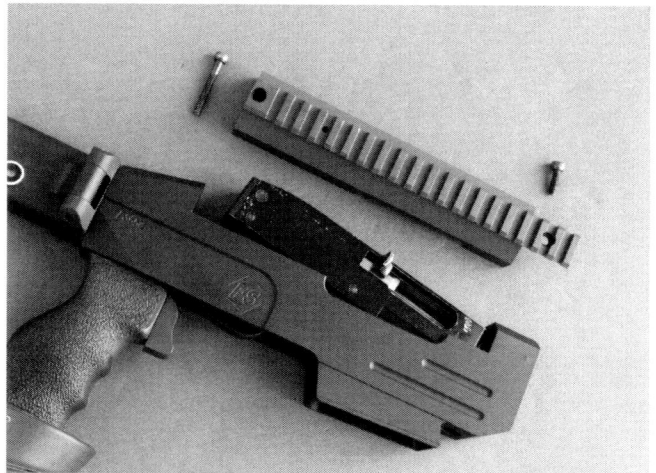

This sequence concludes the stock and pistol grip assembly to the chassis. The second part of the installation process involves assembling of the gun's action on the chassis. To do this, the barrel must first be disassembled from the receiver. Once the barrel and receiver have been separated, perform the following steps.

7. Assemble the receiver on the lower chassis with the front end first and then push down the rear to level the receiver as it is seated into position. At this point, lightly tighten the stock screw at the front-end bottom of the chassis.

Lightly tighten the assembly screw at the bottom once the action has been fully seated to the chassis. Assemble the barrel to the receiver through the opening at the front end of the chassis, complete with the V-block and retaining screws as shown.

8. Insert the barrel complete with its V-block and assembly screws to the opening at the front of the chassis and push in the barrel to its designated assembly hole in the receiver. Prior to barrel assembly, it is best to have the bolt in the open position by using the gun's hold-open device.
9. Tighten the barrel's V-block screws to secure the barrel in the usual manner.

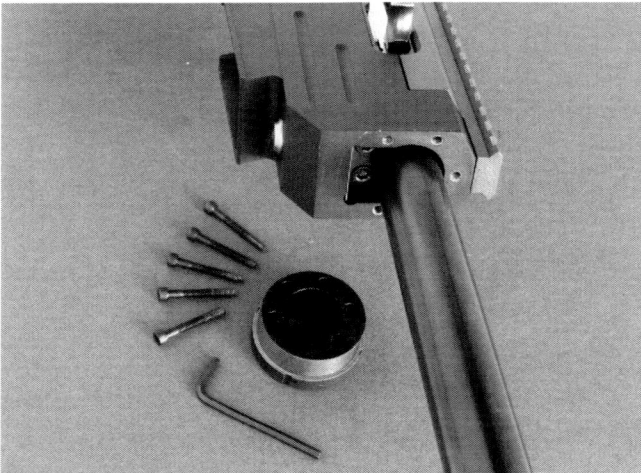

Once the barrel retaining screws are tightened, you may now install the matching forearm adaptor complete with YHM jam nut. The adaptor is secured to the front end of the chassis by five large Allen screws. This is a very solid setup, but it takes time to assemble and disassemble the gun when required.

10. Assemble the forearm adaptor to the front of the lower chassis. Five Allen screws secure the adaptor.

View showing the FS forearm adaptor fully installed complete with YHM jam nut. You will need the YHM forearm wrench to tighten the jam nut properly after the forearm is positioned properly.

11. Assemble the YHM-9433 (jam nut) to the forearm adaptor.

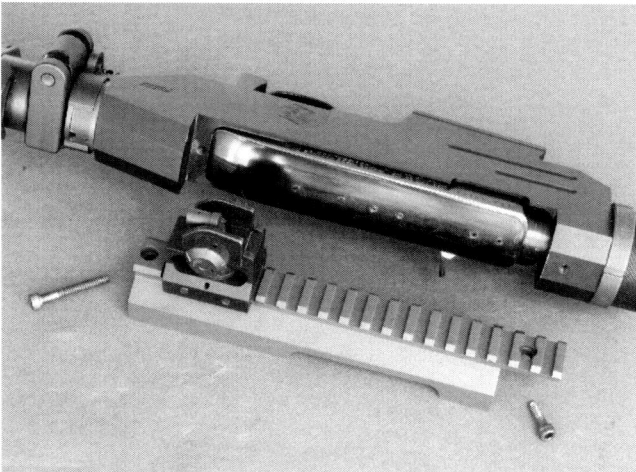

After installation of the forearm adaptor on the FS556 chassis, the matching upper can now be installed. The upper is secured by two large Allen screws on both ends, providing solid mating of both halves.

View showing the FS upper fully seated with rear assembly screw installed. Note the seamless mating of the two halves and very elegant lines of the chassis design.

12. Assemble the FS upper to the lower chassis. The upper and lower chassis are secured together by two mid-sized Allen screws.

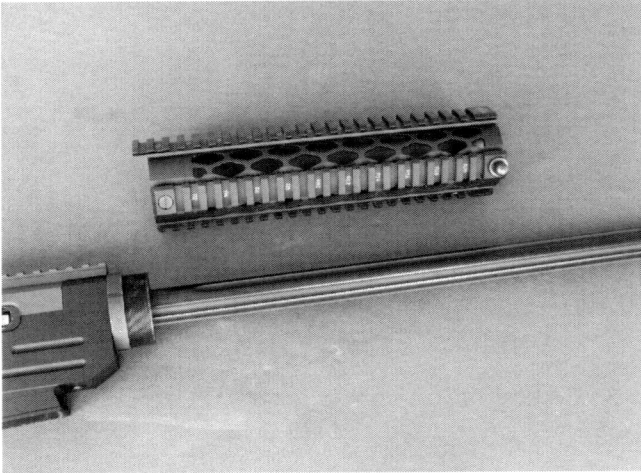

Complementing this setup is the YHM Specter Length Diamond series free-float forearm. I used a Green Mountain 18-inch fluted heavy barrel. Before tightening the jam nut, make sure the top rail of the forearm lines up properly and touches the FS upper rail.

13. Assemble the YHM-9637-DX (Specter forearm) to the adaptor. Rotate the forearm as far as it will go to the rear of the adaptor with the jam nut installed, line up the forearm rail to the rail of the upper chassis, and then tighten the forearm position via the jam nut. You will need YHM-9621 spanner/wrench combination tool to tighten the jam nut properly. This special tool can also be used to tighten the YHM-9483 (barrel nut) as used in the Nordic AR-22 forearm installation. Center the barrel to the forearm and then tighten the screw for the chassis.

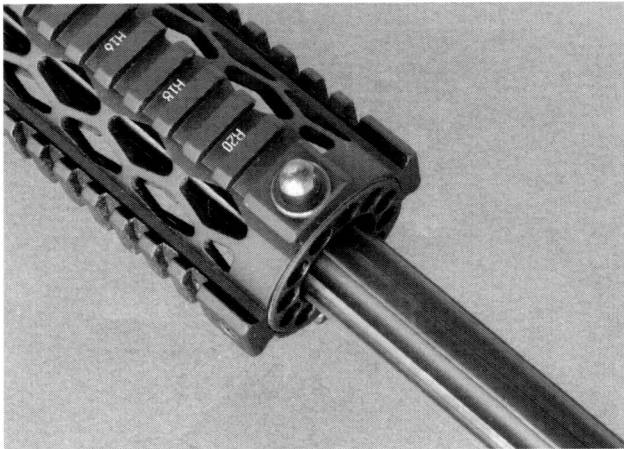

Install the matching YHM endcap (optional), including the sling swivel stud at the prethreaded locations at the front end of the tubing.

Installation of the DPMS four-rail gas block to the GM fluted heavy barrel will require a sleeve improvised from an aluminum pop can to fill in the 0.008-inch gap between the barrel's .920 (OD) and DPMS gas block's .936 (ID). The aluminum sheeting is wrapped around the barrel, and the gas block is slipped over it to attain the desired tightness.

14. Assemble the fake gas tube to the gas block. You can improvise the gas tube from regular 3/16-inch diameter seamless tubing or a mild steel rod about 1 inch long.
15. Tightly roll the aluminum pop can sheet (measuring 4 1/2 x 1 1/2 inches) around the barrel where the gas block will seat and then install the gas block over it to eliminate looseness. This method of assembly applies whenever you are installing a .936 ID gas block to a .920 OD 10/22 barrel.

View showing the four-rail gas block installed halfway past the wraparound shim over the barrel. During this installation procedure, restrain the rolled aluminum sheeting (wrapped tightly around the barrel) at the rear while wiggling the gas block back and forth and pushing the shim in toward the gas block until fully seated. Install the short, fake gas tubing prior to assembling it on barrel as shown.

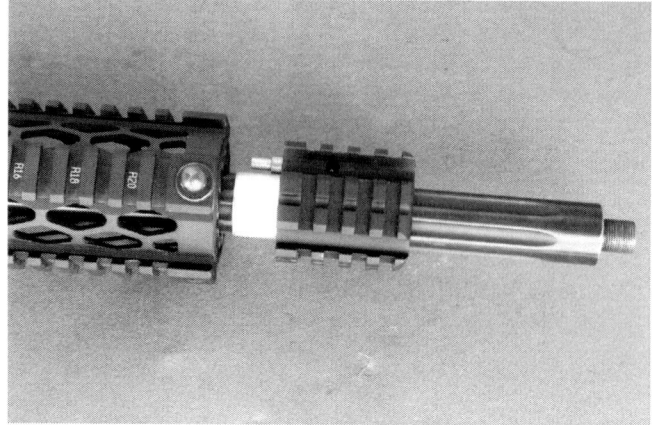

Top view showing the gas block in fully seated position. Make sure to align the top rail of the gas block with the forearm top rail before tightening the set screws of the gas block as shown in order for the front sight to be in line with the matching rear sight.

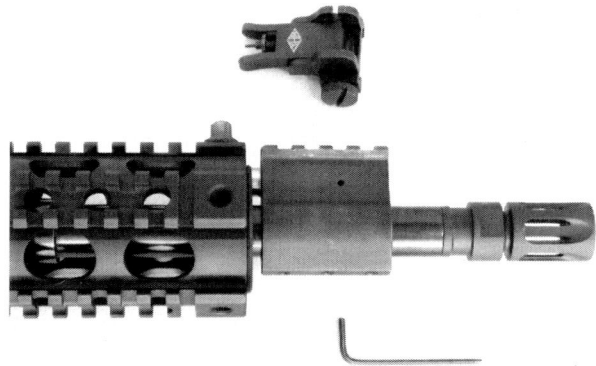

16. Align the hole of the forearm endcap (YHM-9484-B) to the protruding portion of the fake gas tube once the gas block is centered on the top rail of the forearm and push it into its designated position. Leave a gap of approximately .065 inch between the front end of the forearm and the gas block to expose a portion of the fake gas tube for an authentic look. Tighten the set screws (DPMS) or YHM side-mounted assembly screws as needed.

17. Assemble the YHM-9584 (gas block flip-up front sight) to DPMS gas block. If you prefer the YHM-9388 (raised, railed gas block), use the LMT forearm-mounted front sight or YHM-9627 (same-plane, flip front sight) if your preference is for a folding front sight.

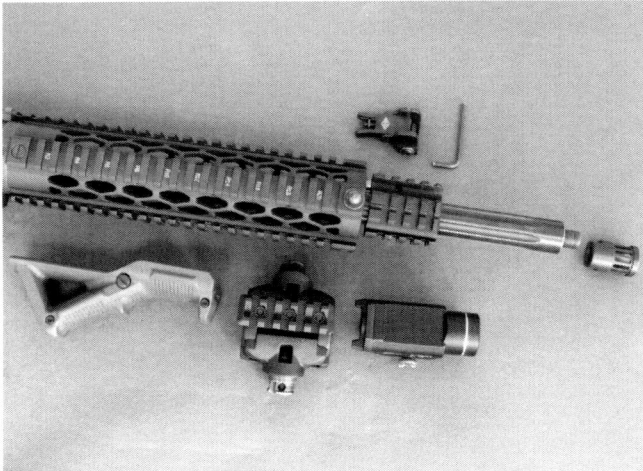

After the assembly of the gas block to the barrel is completed, the YHM folding front sight, Carlson Mini Comp, Magpul AFG foregrip, CAA bipod mount, and TLR-1 weapon light are ready for installation.

View showing the YHM gas block folding front sight, Mini Comp, bipod, and tactical light installed. This is a simple, elegant, and very functional setup that delivers the performance of a world-class tactical 10/22.

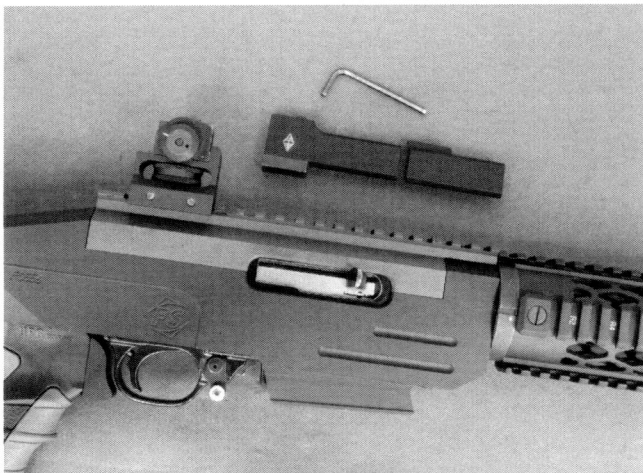

The final step is assembling the YHM A2 rear sight and EOTech riser to the rail of the upper FS chassis. The FS MCR22 is now ready for action.

The fabulous Fabsports Marksman Challenger Rifle (MCR22) fully dressed and ready for action.

18. Install your choice of muzzle brake or compensator device if the barrel is threaded. The Carlson Mini Compensator (available from Brownells) is perfect for this setup because it adds minimal length to the barrel.

19. Install the Magpul MOE 1/2-inch stock riser to the buttstock, then the AFG grip (OD green), CAA bipod, and your choice of laser or light at the lower rail of the forearm.

This concludes the full assembly of the Fabsports Marksman Challenger Rifle system. For disassembly, simply reverse the assembly sequence.

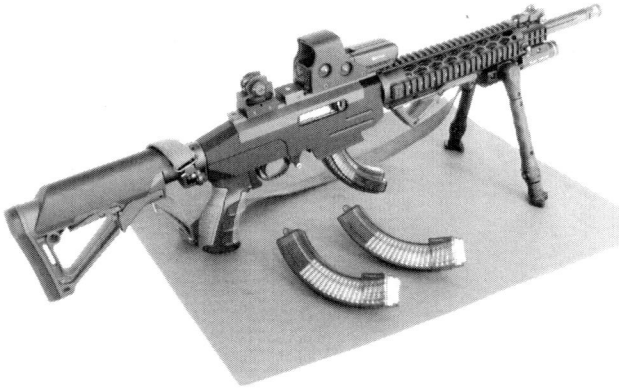

Right-rear view of the MCR22 with EOTech 512 red-dot scope installed. Ram-Line's compact 40-shot magazine (though no longer made) is a perfect addition to this tactical 10/22 package.

Left-side view of the MCR22 with its stock folded. The sling is .30-caliber M1 Carbine accessory.

View showing the EOTech 512 removed from MCR22 to show detail of DPMS tactical rear sight and YHM EOTech riser.

This combination boasts a superlative tactical sighting combination: the YHM A2 rear sight (with optional DPMS hooded peep sight) and EOTech 512 red-dot scope sitting atop the YHM EOTech riser.

STEPS FOR CLEANING, REPAIRING, OR MAINTAINING RIFLE

For regular maintenance, cleaning, or repair that requires access to the action of the gun, follow these procedures.

1. Disassemble the muzzle device (if the gun is so equipped).
2. Disassemble the gas block from the barrel by loosening the screws; leave the front sight in place.
3. Disassemble the forearm from the adaptor. You may need to loosen the jam nut before you can rotate the forearm for removal. You need only to disassemble the bipod legs from its mount at the bottom rail. The bipod mount along with the AFG grip and laser/light can be left installed to the forearm.
4. Leave the YHM jam nut on the FS adaptor and unscrew the five assembly screws to disconnect the adaptor from the lower chassis.
5. Disassemble the V-block screws to free the barrel from the receiver.
6. Unscrew the stock screw to disconnect the receiver from the lower chassis.
7. Separate the chassis from the receiver by lifting its rear section first to clear its connection with the chassis. The action is now clear of the chassis and ready to be serviced or cleaned in the usual manner.

For reassembly, simple reverse the above sequences.

CHAPTER 14

Variations: Alternative Precision Rifle Profiles

Have you ever wondered how tactical firearm producers can introduce multiple models from the same gun? The answer is quite simple: accessorizing. Take, for example, a typical M4 Carbine. A manufacturer can easily tailor numerous weapon profiles out of the same-action barrel by simply installing different types and styles of stocks, pistol grips, forearms, backup iron sights, and muzzle devices, and then giving each variant a new model designation. Virtually all military and sporting arms manufacturers have applied this simple and efficient formula with great success. It is foolproof and economical since the manufacturers don't have to spend time and money designing new fixtures and tooling for the production of furniture and other major components that can be readily acquired from independent sources. The manufacturers simply select accessories that can be matched with their own production parts.

Building your own customized tactical 10/22 is no different. By now, you should have a clear understanding of how to build a reasonably good tactical 10/22 the way the professionals do. The primary pitfall of would-be tactical gun builders is failing to have a clear view of what an ideal tactical gun is all about. Many of the homemade 10/22 rail guns being bragged about by hobbyist gunsmiths and basement experimenters lack the appeal and quality of a professionally crafted gun. They are often overdecorated with cheap or expensive accessories (or a combination of both) that simply do not correlate with each other. This kind of setup results in a crude, unwieldy,

heavy, and unattractive tactical gun that nobody likes. The whole objective is to blend aesthetics and ergonomics, and back them up with reliability and accuracy, to produce the perfect gun for you.

Sadly, many of the old masters of weapons design are slowly fading away from the gun scene. For the few who are still around, there is no greater accomplishment than sharing the experience they have learned over the years with a new generation of craftsmen who are eager to learn the trade and continue this revered American tradition so envied the world over. The primary goal of this book is to guide the aspiring amateur gunsmith or hobbyist in the creation of his own tactical gun, but perhaps in the process a few will discover that they have a special talent for this type of work and join the new breed of firearms designers. Who knows? It could even be you.

This chapter features additional variations of tactical .22 rifles for your customizing reference. They use the same barreled action and custom platforms made by RB, Nordic, and Fabsports (as outlined in Chapters 11, 12, and 13) but utilize different furniture, pistol grips, backup sights, optics, and rails systems. You will notice that among the three platforms used in these custom gun projects, the Nordic AR-22 has the most weapon profiles presented, proving its amazing versatility over its competitors. Use the platforms as a basis for creating your own versions of the tactical 10/22 or improve on them as you see fit.

I have broken down each chassis systems into three basic sections to make it simpler to understand how each additional weapon profile is put together. I could build even more weapon configurations with the same accessories I have on hand and come up with a dozen more models quite easily, proving the unlimited applications of these ultramodern tactical gadgets. As you go through the process of building your own tactical guns, you will gain more confidence and experience not only in envisioning advanced tactical weapons concepts of your own but also, and more importantly, how to put the design together professionally.

Throughout this book, I have presented what I consider the most sophisticated 10/22 tactical rimfire weapon systems based on decades of personal experience and research. In reality, the only way to truly evaluate a first-rate custom tactical weapon (no matter what type or caliber it is) is to have it in your hands, holding it and putting it through its paces. This usually means that you must have a pile of parts and accessories that you can experiment with in building numerous weapon profiles to find out which setup works best for you. Unfortunately, that is a costly undertaking for the average consumer, especially during our present economic times. The good news is, you don't have to do this yourself, because I have already done that for you.

Although bringing down the cost as much as possible is my primary goal when building my own custom guns, using the cheapest components available proved to be counterproductive in many instances. Therefore, the most logical choice is opting for high-quality, mid-priced parts and accessories that are just as good as the more expensive high-end brands. With this in mind, expect to invest a few dollars more in building your own tactical .22 than you would spend on a mass-produced, factory-made model. On the positive side, take into consideration that the custom guns presented here feature CNC-machined, aircraft-grade chassis (not die-cast metals or plastics, as used on some of the factory-made clones). All furniture pieces used are genuine AR-15 custom-grade components, as are the backup iron sights and mounting rails. No shortcuts and no Mickey Mouse parts are used here. You are building your pride and joy. Backed by the 10/22's legendary

reliability and never-ending flow of custom-made parts and accessories for your future upgrading requirements, your custom rifle will be one that you can be proud to own, compete with, or show off. If you keep it in good condition, its value will only appreciate over time.

After going through the basic assembly instructions of various weapon profiles as outlined in Chapters 11, 12, and 13, you should be familiar with the techniques on how to build custom tactical 10/22s the way the professionals do. By simply reviewing the additional photographs in this chapter, you will have a clear idea on how to put these guns together and what parts, accessories, and tools you are going to need. This section is meant to test how much you have learned so far.

Weapon configurations vary slightly from one to the other. Pick the one you like and build the gun or guns for your own enjoyment and pride. Keep the faith, stay safe, and good luck with your new gun projects.

RB PRECISION ADDITIONAL WEAPON PROFILES

The UMR22 Mk-1, a variation of the UMR match rifle system utilizing CAA furniture in two-tone format, complemented by the Mitchell 50-shot teardrop magazine. The Mk-1 was set up to accommodate both the Keng's Versa-Pod (installed) and CAA side-mounted bipod (seen with legs detached).

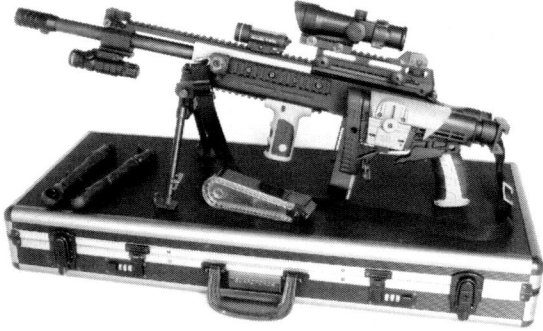

The UMR Mk-1 with its stock in folded position. Backup iron sights are RRA tactical carry handle A2 rear sight and Brownells tactical flip front sight. The optic is a Bushnell Trophy red-dot scope.

My UMR Mk-2 featuring the AGP polymer, side-folding stock, seen here in folded position. Ideally, the side folder works on the left side so as not to interfere with the cocking handle and ejection port.

Left-side view of the UMR Mk-2 with its stock in extended position. This 10/22 tactical black rifle features Magpul grips, Leapers SWAT tactical full-size scope, and CAA side-mounted bipod.

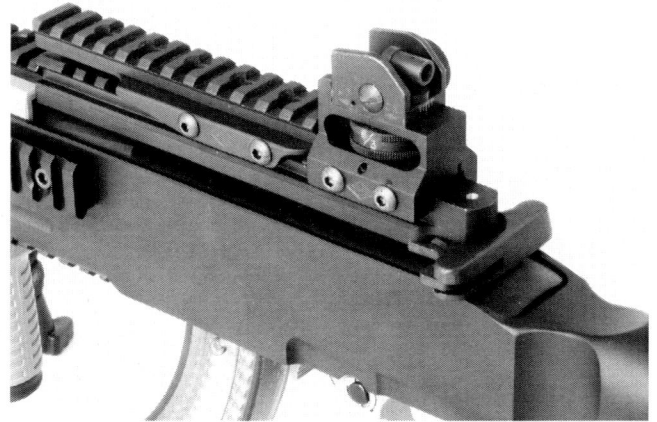

Close-up of the UMR Mk-1 nonfunctional AR-15 charging handle. Note the YHM A2-style rear sight and YHM 5-inch rail extension used as a rail riser in this setup.

NORDIC ADDITIONAL WEAPON PROFILES

My XM122 is a rimfire superclone of the famed XM177 Commando from the 1980s. This tactical rimfire package features the AGP side-folding stock and Cavalry Arms' slimline C3 two-piece plastic hand guard. Shown with the gun are alternate fake suppressors made by Spike's Tactical (on gun), Tactical Innovations, and Fabsports (with perforations).

Right-side view of the XM122 with its stock extended. Backup iron sights are DPMS tactical rear sight and Brownells M4 flip-up front sight, complemented by a compact Bushnell HOLO-Sight red-dot scope. High-capacity magazines are Ram-Line (banana) and Black Dog (drum) innovations.

View of the XM122 with the SD-1 fake suppressor removed to show the concealed portion of the barrel when the shroud-type device is installed. The YHM Phantom compensator is used as an alternate muzzle device in this setup. The Harris bipod completes the classic Commando format.

Top-rear view of the XM122 Commando.

Left-front view of the XM122 Commando. This rimfire tactical marvel looks good from every angle and has the reliability and accuracy to back up its looks.

I designated this the Tactical Target Rifle (TTR-22). This lightweight target rifle features the AGP folding stock and the CAA UPG16 grip with optional target stand complemented by a Magpul AFG front grip.

Right-side view of the TTR-22 with its stock folded, which blocks both the ejection port and cocking handle. To operate the gun, the stock must be in the extended position. The CAA side-mounted tactical bipod complements the Leapers full-size SWAT scope for long-range shooting.

The Black Knight is a variation of the TTR-22 "sniper" format. It features a CAA/EMA Tactical buttstock and CGRIP forward grip complemented by a DPMS Panther Tactical Grip. Note the high-performance Smith Enterprise AR-15/M16 Vortex flash eliminator.

Rear-right view of the Black Knight, shown with alternate muzzle devices and high-capacity magazines, backup iron sights, scope riser, and YHM rifle-length four-rail free-float forearm. The CAA side-mounted bipod provides a stable platform for long-range shooting.

The Sturmgewehr 22 (STG22) is a variation of the PSG22 that incorporates a two-piece plastic hand guard instead of the one-piece aluminum free-float type. The furniture used in this ultrasophisticated setup is by Magpul.

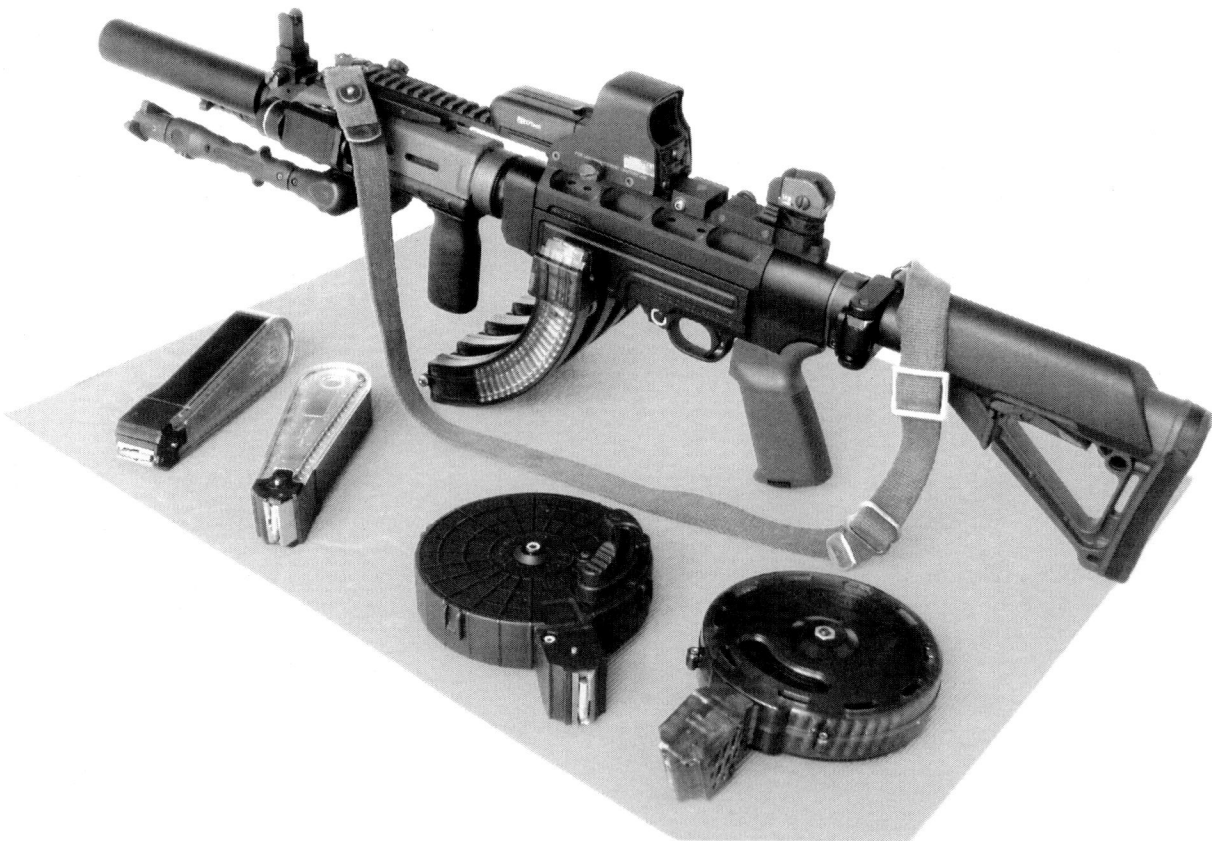

Left-side view of the STG22 shown with an assortment of 50-shot magazines in banana (Ram-Line), teardrop (Mitchell Arms), and 50-shot drum configurations (Black Dog and Pro-Mag).

The STG22 in long-range rifle format. The YHM EOTech riser was replaced with Brownells AR-15 flattop riser to accommodate the full-size Leapers SWAT scope mated to RRA 30mm high-rise cantilever scope mount. For best shooting comfort in a high-mounted optic, install the Magpul 1/2-inch cheek riser to the CTR stock.

The STG22 with its stock in the folded position.

The STG22 Model 2 with YHM Diamond Series carbine, free-float aluminum forearm in all-black Magpul attire.

Left-side view of the STG22 Model 2 with its stock in the folded position.

The superb Magpul CTR buttstock is designed to accept optional accessories to enhance its versatility, including the cheek riser (available in 1/4-, 1/2-, and 3/4-inch height) and the enhanced recoil pad.

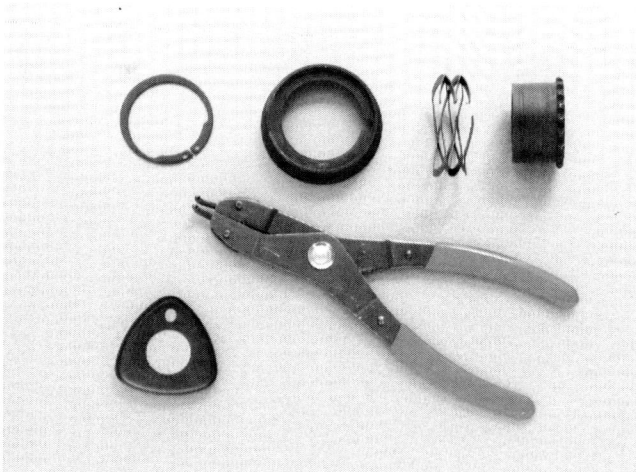

Installation of a two-piece hand guard to the AR-22 chassis will require the AR-15 Delta ring/barrel nut assembly and hand guard cap shown here in strip view. You will need a pair of snap ring pliers to attach the Delta ring to the barrel nut.

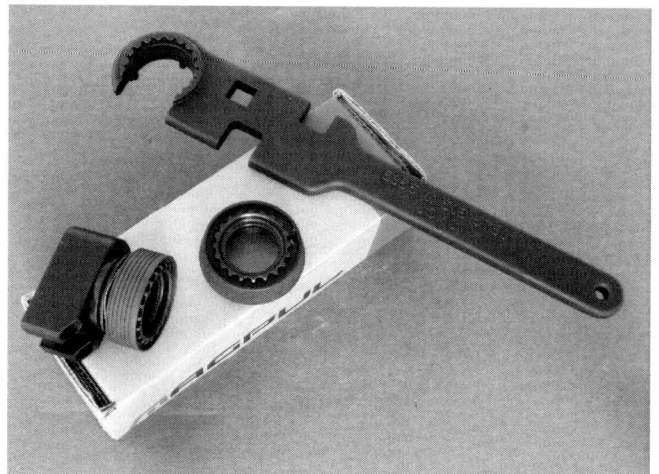

The installation of the Delta ring/barrel nut assembly to the AR-22's forearm adaptor will require an armorer's wrench, such as the Smith Enterprise tool seen in this photo. In the center is a spare Delta ring/barrel nut assembly in tapered format. The Delta ring on the forearm adaptor is the vintage straight format.

View showing the fully assembled Delta ring/barrel nut assembly prior to its installation on the forearm adaptor.

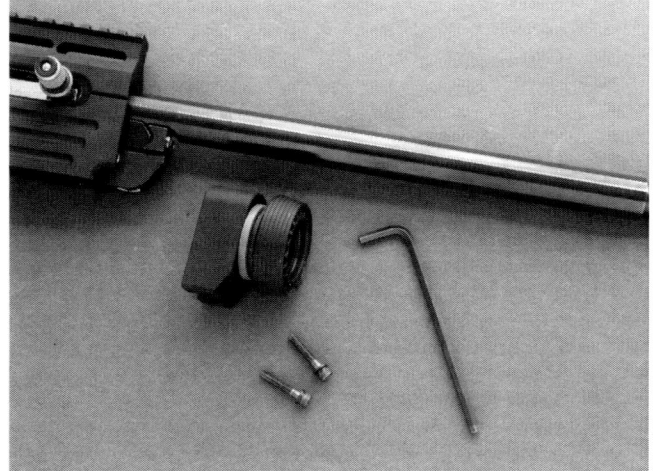

View showing the Delta ring/barrel nut assembly installed to the forearm adaptor. I added a plastic spacer ring behind the barrel nut to cover the unsightly gap between the adaptor and Delta ring. The plastic spacer ring can be improvised from an old vacuum cleaner extension.

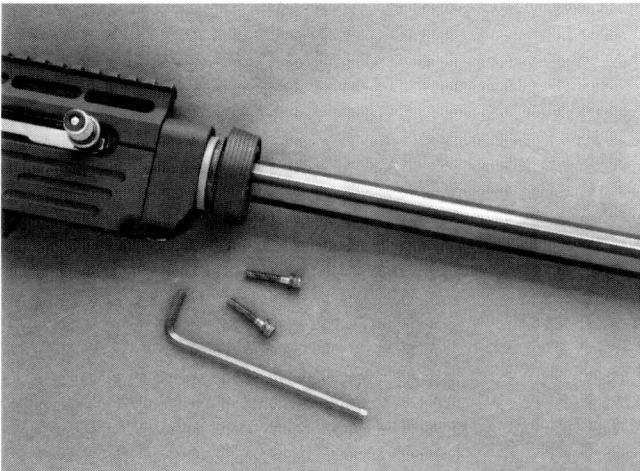

View showing the forearm adaptor complete with Delta ring/barrel nut assembly positioned at the front end of the chassis, ready to be secured by the assembly screws.

With the right accessories, a two-piece plastic hand guard with side, top, and bottom rails will accommodate the same type of accessories as a one-piece free-float forearm (seen above) will take.

The two-piece STG22 plastic hand guard is assembled in the same manner as you would in a regular AR-15/M16 rifle to the Delta ring/barrel nut assembly mounted on the AR-22 forearm adaptor. The front end of the hand guard is secured by the matching cap seen on the left.

AR-15 hand guard caps are made in a round or triangular format. Seen in the photo is the round type used on carbine models with standard .750 OD barrels (left). The one on the right has the center hole enlarged to accommodate the .920 OD heavy barrel used in the 1022.

View showing the various components to complete the hand guard, front sight, and muzzle device group of the STG22.

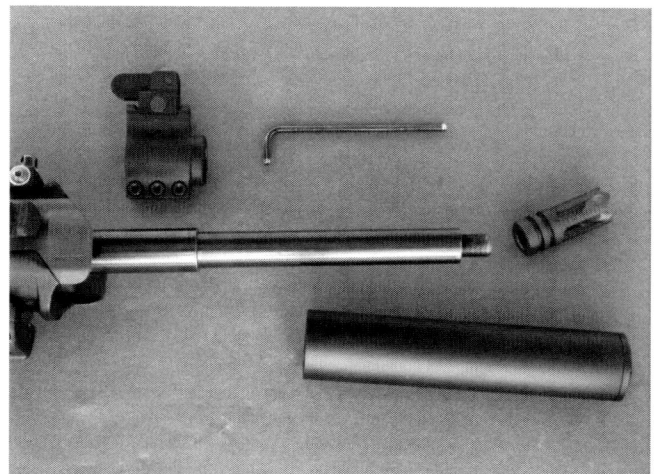

Once the hand guard endcap has been assembled in place, the YHM same-plane, railed, bull-barrel gas block (top) is pushed against the endcap to tighten the assembly of the hand guard before the gas block screws are tightened.

View showing the gas block and fake suppressor installed. You may have to improvise a tool (shortened Allen wrench) to get to the hidden screw inside the front end of the hand guard.

Close-up view showing the shortened Allen wrench tightening the hidden screw behind the front end wall of the hand guard.

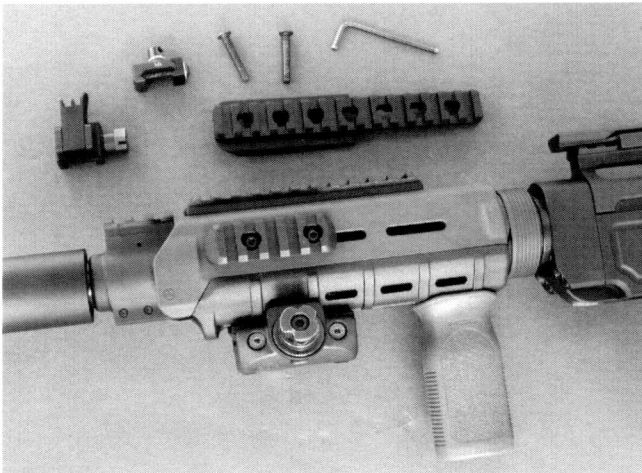

Reverse-assemble the YHM 6-inch rail extension to the Magpul rail on top of the hand guard until it meets the EOTech riser front end to form the uninterrupted rail system of this setup.

The AR-22 A2 features the classic styling of a full-size AR-15 incorporating the CAA Carry Handle (CH) sight and matching DPMS detachable front sight. Alternate backup sights and muzzle device are shown below the gun.

The AR-22 A2 with its stock in the folded position. The optional adjustable cheekpiece on the CAA CBS stock is perfect when mounting an optic over a carry-handle style sight, such as the RRA accessory shown in this photo. Note the reversed-mounted YHM 5-inch rail extension above the RRA rail used to clear the rear sight protective wings when a full-size scope is installed.

My custom Specter tactical 10/22 features two-tone (dark earth and black) Magpul furniture and a Dlask 18-inch match barrel complemented by YHM mid-length Diamond Series free-float four-rail forearm. The Fabsports fake suppressor in dark earth color (lower right) perfectly matches the furniture. The rifle has the DPMS Meculek compensator installed.

Left-side view of the Specter with the fake FS suppressor installed. The tan-colored sling is a vintage Galil .22 accessory. This gun defines the true meaning of a tactical .22 show-stopper. It is what separates a custom-built gun from a mass-produced model.

Another superb tactical 10/22 setup using the AR-22 chassis is the Tracker model. The Tracker features Magpul pistol grips and two-piece, mid-length, plastic hand guards complemented by an RRA entry buttstock. Backup iron sights are by RRA and Brownells.

Left-side view of the Tracker with the stock in the folded position. The compact, fixed-type RRA plastic stock is light and strong, and it provides an excellent cheek rest. The use of plastic furniture and a heavy barrel creates a perfectly balanced all-around rifle for the busy marksman.

Left-side view of the Ruger SR22 featuring Tapco SAW-style pistol grip and Fusion forward grips in dark earth. Complementing the grips is the ACE M4-style buttstock complete with CAA stock saddle.

Right-side view of the SR22 in classic AR rifle format, complemented by CAA Carry Handle sight and matching front sight. Note the SOG Armory graphite vertical grip and YHM railed (raised) bull barrel gas block (which will require a sleeve for an .0750 OD barrel to mount the gas-block front sight).

Comparison between my custom AR-22 tactical carbine (bottom) and its SR22 counterpart in traditional AR-15 styling.

The upper half of the SR22 chassis features a rail system that can accommodate various tactical accessories, such as backup iron sights and optics. Unlike other production tactical .22s on the market, the SR22 has a lot of room for customizing and upgrading for hobbyist gunsmiths and tactical gun aficionados.

FABSPORTS FS556 ADDITIONAL WEAPON PROFILES

The Adventurer is a colorful tactical 10/22 setup using the FS556 chassis complemented by Magpul's dark earth plastic furniture and backup iron sights. This gun uses the DPMS .936 ID four-rail gas block. The YHM gas block riser will be needed here to raise the height of the gas block at top in order for the lower front sight to co-witness with the rear sight.

Left-front view of the Adventurer with the stock in folded position. This rifle features the Double Star AR-15 mini-compensator. The full-size scope rides on a YHM scope riser.

I call this model the FS Tactical Sniper Simulator (TSS22). It features a unique blend of AR-15 furniture made by Magpul (CTR stock with 3/4-inch cheek riser), DPMS (Panther grip), and CAA CGRIP (forward grip).

The TSS22 with the stock folded. An Eagle International 30-shot banana magazine complements this high-tech setup.

The TSS22 with alternate buttstocks, complete with cheek risers. From left to right: M4 stock with EMA tactical CP2 cheek riser, Magpul CTR with 3/4-inch cheek riser, and ACE FX tubular stock with CAA FALC cheek riser. On the gun is the CAA CBS stock with adjustable cheekpiece.

The FS556 Masterpiece (TM22) is Fabsports tactical 10/22. This beautiful gun features two-tone (tan and black) furniture complemented by the YHM Diamond Series Specter free-float aluminum forearm.

The FS556 Masterpiece with a host of accessories that can be employed by the operator to meet specific requirements. Note the SIG SAUER mini-laser mounted on the side rail of the forearm. The unique flash hider with helical slots is made by North Eastern Arms in Canada.

Rear-right-side view of the Masterpiece tactical 10/22 rifle. Backup iron sights are DPMS tactical rear sight and YHM flip front sight with bottom rails.

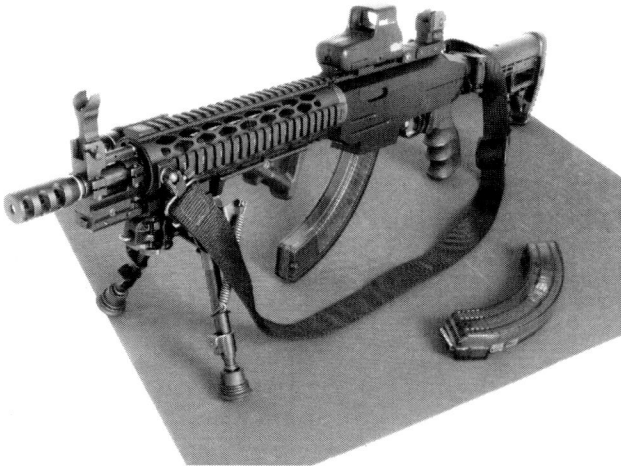

Left-side view of the FS556 Masterpiece in all-black attire, utilizing CAA/EMA Tactical furniture. Note the skeletonized CAA buttstock and finger-grooved pistol grip (G16), complemented by a Magpul AFG2 forward grip.

Right-side view of the FS556 Masterpiece with the laser mounted to the bottom rail of the gas block for more precise target acquisition, as opposed to mounting the laser to the forearm, which is not always perfectly centered on the barrel axis. The tactical light is mounted on the side rail.

This unique FS556 tactical carbine setup features the CAA Carry Handle (CH) sight, complemented by Midwest Industries' A2 adjustable cantilever mount. The cantilever mount positions the optic forward of the carry handle (instead of over the handle), thus eliminating the need for a cheek riser for correct shooting posture.

Left-side view of the FS556 tactical 10/22 carbine with the stock folded. This two-tone beauty (olive drab and black) is equipped with the Vltor VC-1 flash suppressor and EOTech 512 red-dot scope. The gas block is the Brownells AR-15 modular design with side and bottom rails removed. The YHM same-plane fixed-front sight will require the YHM gas block riser to elevate the position of the front sight so that it will co-witness with the rear sight.

Every tactical rimfire aficionado dreams of creating the ultimate gun, complete with all the bells and whistles. Fulfilling that dream has never been easier to accomplish, with the vast array of tactical gun accessories available for the Ruger 10/22. With a little know-how and imagination, the sky is the limit. Always remember, a true tactical firearm, no matter what caliber it is, must be attractive, user friendly, reliable, and deadly accurate.

CHAPTER 15

Sources for Parts and Accessories

BROWNELLS
200 South Front Street
Montezuma, IA 50171
www.brownells.com
1-800-747-0015
For more than 30 years, I have been actively designing and building prototype firearms and product improvements for various arms manufacturers, both in the United States and abroad. My primary source for gun parts, accessories, finishes, tools, and everything else is Brownells. This company is the undisputed leader among high-volume suppliers of top-quality firearms products and gunsmithing tools for manufacturers, gunsmiths, and hobbyists the world over. Brownells' extensive product line caters to virtually everyone's budget; plus it has no minimum orders and all sales are backed by a no-hassle, 100-percent-satisfaction guarantee. Most of the parts and accessories used in creating the high-tech tactical 10/22s featured in this book are available directly from Brownells.

Specific parts or accessories that may not be available from Brownells can be purchased directly from the manufacturers.

ACE, LTD., USA
P.O. Box 430
Winchester, KY 40391
888-736-7725
www.riflestocks.com

AGP ARMS, INC.
1930 East 3rd Street, Suite 12
Tempe, AZ 85281
480-983-6083
www.agparms.com

COMMAND ARMS ACCESSORIES/ EMA/CAA TACTICAL
1208 Branagan Drive
Tullytown, PA 19007
215-949-9944
www.commandarms.com

DLASK ARMS
#16 7167 Vantage Way
Delta, British Columbia
Canada V4G-1K7
604-952-0837
www.dlaskarms.com

FABSPORTS
7191 Beaubien Est.,
H1M-2Y2
Montreal, Quebec
Canada
514-355-3423
www.fabsports.ca
Fabsports supplies tactical weapon accessories and 10/22 custom chassis in Canada.

LEWIS MACHINE AND TOOL COMPANY
1305 11th Street West
Milan, IL 61264
309-787-7151
www.lewismachine.net

MAGPUL INDUSTRIES CORPORATION
P.O. Box 17697
Boulder, CO 80308
877-462-4785
www.magpul.com

MISSION FIRST TACTICAL
780 Haunted Lane
Bensalem, PA 19020
267-803-1517
www.missionfirsttactical.com

NORTH EASTERN ARMS (NEA)
1399 Portage Road, R1
Kirkfield, Ontario
Canada KOM-2BO
705-879-2870
www.northeasternarms.com

NORDIC COMPONENTS
1156 Highway 7 East
P.O. Box 429
Hutchinson, MN 55350
877-549-9893
www.nordiccomp.com

RB PRECISION
518 Forrest Road
East Moline, IL 61244
309-281-1119
www.rbprecision.com

ROCK RIVER ARMS
1042 Cleveland Road
Colona, IL 61241
866-980-7625
www.rockriverarms.com

STURM, RUGER & CO., INC.
411 Sunapee Street
Newport, NH 03773
603-865-2442
www.ruger.com

TECH-SIGHTS, LLC
904 Deer Run Drive
Hartsville, SC 29550
843-332-8222
www.tech-sights.com

YANKEE HILL MACHINES
20 Ladd Avenue, Suite 1
Florence, MA 01062
877-892-6533
www.yhm.net